Developing
sustainable
food value chains

Guiding principles

David Neven

FOOD AND AGRICULTURE ORGANIZATION
OF THE UNITED NATIONS
Rome, 2014

RECOMMENDED CITATION
FAO. 2014. *Developing sustainable food value chains – Guiding principles.* Rome

Cover photos (from top)
Women work the rice fields in Vietnam and most developing countries. Lothar Wedekind/IAEA
A worker at the Valdafrique Processing Plant. ©FAO/Seyllou Diallo
Employee canning artichoke hearts at the AgroMantaro processing plant in Junín, Peru. ©FAO/David Neven
Women buying fish from the wholesaler. ©FAO/Antonello Proto
A little girl eats a freshly-made roti. S. Mojumder/Drik/CIMMYT

The designations employed and the presentation of material in this information product do not imply the expression of any opinion whatsoever on the part of the Food and Agriculture Organization of the United Nations (FAO) concerning the legal or development status of any country, territory, city or area or of its authorities, or concerning the delimitation of its frontiers or boundaries. The mention of specific companies or products of manufacturers, whether or not these have been patented, does not imply that these have been endorsed or recommended by FAO in preference to others of a similar nature that are not mentioned.

The views expressed in this information product are those of the author(s) and do not necessarily reflect the views or policies of FAO.

ISBN 978-92-5-108481-6 (print)
E-ISBN 978-92-5-108482-3 (PDF)

© FAO, 2014

FAO encourages the use, reproduction and dissemination of material in this information product. Except where otherwise indicated, material may be copied, downloaded and printed for private study, research and teaching purposes, or for use in non-commercial products or services, provided that appropriate acknowledgement of FAO as the source and copyright holder is given and that FAO's endorsement of users' views, products or services is not implied in any way.

All requests for translation and adaptation rights, and for resale and other commercial use rights should be made via www.fao.org/contact-us/licence-request or addressed to copyright@fao.org.

FAO information products are available on the FAO website (www.fao.org/publications) and can be purchased through publications-sales@fao.org.

Contents

Foreword	v
Acknowledgements	vi
Executive summary	vii
Abstract	x
About the author	xi
Acronyms	xii

Chapter 1
Introduction — 1

Chapter 2
Concept and framework — 5
   2.1 » Defining the concept — 6
   2.2 » Related concepts — 8
   2.3 » The sustainable food value chain framework — 10

Chapter 3
The sustainable food value chain development paradigm — 13
   3.1 » Return on assets: farming as a business and development of small and medium-sized agro-enterprises — 14
   3.2 » Salary income: creating decent work opportunities — 16
   3.3 » Taxes: ecosocial progress — 16
   3.4 » Food chain fallacies — 17
   3.5 » The sustainable food value chain development paradigm: conclusion — 20

Chapter 4
Ten principles in sustainable food value chain development — 21
   4.1 » Measuring performance of food value chains – sustainability principles — 23
   4.2 » Understanding food value chain performance – analytical principles — 35
   4.3 » Improving food value chain performance – design principles — 46

Chapter 5
Potential and limitations — 59

Chapter 6
Conclusions — 63

References — 65

Annex
Concepts related to the value-chain concept — 71

Figures

1. A breakdown of the concept of value added — 7
2. The sustainable food-value-chain framework — 11
3. The sustainable food-value-chain development paradigm — 15
4. Principles of sustainable food-value-chain development — 23
5. Sustainability in food-value-chain development — 24
6. Positive feedback loop driving sustained growth — 27
7. Examples of constraints and leverage points in value chains — 37

Boxes

Illustration of Principle 1: The potato value chain in India — 28
Illustration of Principle 2: The pineapple value chain in Ghana — 31
Illustration of Principle 3: The beef value chain in Namibia — 34
Illustration of Principle 4: The vegetables value chain in the Philippines — 38
Illustration of Principle 5: The tea value chain in Kenya — 41
Illustration of Principle 6: The rice value chain in Senegal — 45
Illustration of Principle 7: The coffee value chain in Central America — 49
Illustration of Principle 8: The ndagala value chain in Burundi — 51
Illustration of Principle 9: The dairy value chain in Afghanistan — 53
Illustration of Principle 10: The salmon value chain in Chile — 56

Foreword

The development of sustainable food value chains can offer important pathways out of poverty for the millions of poor households in developing countries. Food value chains are complex systems. The real causes for their observed underperformance may not always be obvious. Typically, multiple challenges have to be tackled simultaneously in order to truly break poverty cycles. This in turn implies the need for collaboration among the various stakeholders in a value chain, including farmers, agribusinesses, governments and civil society. Further compounding the challenge, improvements to the value chain must be economically, socially and environmentally sustainable: the so-called triple bottom line of profit, people and planet.

Around the world, development practitioners in public, private and non-governmental organizations are constantly designing and implementing innovative solutions to address these challenges. These practitioners facilitate the upgrading of products, technologies, business models, policy environments and so on. Some of these solutions fail to have a lasting impact, while others succeed in improving the system at scale and in a sustainable manner. Either way, valuable lessons are learned.

In its role as a global knowledge broker aiming to enable the development of inclusive and efficient agricultural and food systems, the Food and Agriculture Organization of the United Nations (FAO) has initiated a new set of handbooks to capture and disseminate the lessons learned from these experiences. This handbook is the first in the set and it contributes to the achievement of FAO's Strategic Objective Four: *Enable inclusive and efficient agricultural and food systems*. It sets out the overall framework and a set of principles to guide sustainable food value chain development in practice. Subsequent handbooks on this core theme will focus on particular aspects of the approach, taking a systems perspective to present both the main challenges and some of the most promising ways to resolve them.

It is expected that this new handbook, and the ones which will follow, will facilitate the spread among practitioners of new ideas and knowledge related to the development of sustainable food value chains. If successful, it is hoped that this will lead to greater, faster and more lasting impacts in terms of growth in profitability of agribusiness and farming, creation of decent employment, generation of public revenue, strengthening of the food supply and improvement in the natural environment.

Eugenia Serova
Director
Rural Infrastructure and Agro-Industries Division,
FAO, Rome

Acknowledgements

This handbook benefited greatly from the contributions of many people.

First and foremost, special thanks goes to the team who supported the development of the handbook throughout its entire process: Martin Hilmi (Rural Infrastructure and Agro-Industries Division [AGS], FAO), Giang Duong (AGS, FAO) and Victor Prada (AGS, FAO).

Thanks are extended to those who reviewed earlier drafts or contributed insights, illustrations, or definitions. They include: Heiko Bammann (AGS, FAO), Florence Tartanac (AGS, FAO), David Hitchcock (AGS, FAO), Siobhan Kelly (AGS, FAO), Pilar Santacoloma (AGS, FAO), Emilio Hernandez (AGS, FAO), Djibril Drame (AGS, FAO), Eugenia Serova (AGS, FAO), Guy Evers (Investment Centre Division, Africa Service [TCIA], FAO), Claudio Gregorio (Investment Centre Division, Near East, North Africa, Europe, Central and South Asia Service [TCIN], FAO), Gunther Feiler (Investment Centre Division, Technical Cooperation Department [TCID], FAO), Astrid Agostini (TCID, FAO), Luis Dias Pereira (Investment Centre Division, Latin America, the Caribbean, East Asia and the Pacific Service [TCIO], FAO), Lisa Paglietti (TCIA, FAO), Dino Francescutti (TCIO, FAO), Emmanuel Hidier (TCIN, FAO), Jeanne Downing (United States Agency for International Development), Andrew Shepherd (Technical Centre for Agricultural and Rural Cooperation), Bill Grant (DAI, Bethesda, MD, UNITED STATES), Mark Lundy (International Center for Tropical Agriculture [CIAT]), Jean-Marie Codron (French National Institute for Agricultural Research [INRA]), Paule Moustier (French Agricultural Research Centre for International Development [CIRAD]), Etienne Montaigne (INRA), Cornelia Dröge (Eli Broad College of Business, Michigan State University [MSU], MI, UNITED STATES), Bixby Cooper (Eli Broad College of Business, MSU), Gerhard Schiefer (University of Bonn, Germany), Eriko Ishikawa (International Finance Corporation [IFC]), Alexis Geaneotes (IFC), Andriy Yarmak (Investment Centre Division [TCI], FAO), Michael Marx (TCI, FAO), Kristin De Ridder (consultant). Special thanks go to Matty Demont (International Rice Research Institute), Jorge Fonseca (AGS, FAO) and Nuno Santos (Investment Centre Division, Europe, Central Asia, Near East, North Africa, Latin America and the Caribbean Service, FAO) for the peer-review of the final draft.

Finally, I wish to thank Larissa D'Aquilio for coordinating the publication production process, Simone Morini for the layout and cover design, Paul Neate for the copy editing and Lynette Chalk for the proofreading.

Executive summary

Over the last decade, the value chain (VC) has established itself as one of the main paradigms in development thinking and practice. This is why the Food and Agriculture Organization of the United Nations (FAO) has launched this new set of handbooks on sustainable food value chain development (SFVCD), and this is the first in the set. These handbooks aim to provide practical guidance on SFVCD by facilitating the spread of innovative solutions emerging from the field to a target audience of policy-makers, project designers and field practitioners.

This first handbook provides a solid conceptual foundation on which to build the subsequent handbooks. It (1) clearly defines the concept of a sustainable food value chain; (2) presents and discusses a development paradigm that integrates the multidimensional concepts of sustainability and value added; (3) presents, discusses and illustrates ten principles that underlie SFVCD; and (4) discusses the potential and limitations of using the VC concept in food systems development.

DEFINING A SUSTAINABLE FOOD VALUE CHAIN

For the purposes of this publication, a sustainable food value chain (SFVC) is defined as:

> *the full range of farms and firms and their successive coordinated value-adding activities that produce particular raw agricultural materials and transform them into particular food products that are sold to final consumers and disposed of after use, in a manner that is profitable throughout, has broad-based benefits for society, and does not permanently deplete natural resources.*

Unlike related concepts, such as the *filière* and the supply chain, the SFVC concept simultaneously stresses the importance of three elements. First, it recognizes that VCs are dynamic, market-driven systems in which vertical coordination (governance) is the central dimension. Second, the concept is applied in a broad way, typically covering a country's entire product subsector (e.g. beef, maize or salmon). Third, value added and sustainability are explicit, multidimensional performance measures, assessed at the aggregated level.

THE SUSTAINABLE FOOD VALUE CHAIN DEVELOPMENT PARADIGM

The SFVCD paradigm starts from the premise that food insecurity is foremost a symptom of poverty. Households that have sufficient financial resources at all times create the effective demand that drives the supply of food. On the supply side, competitive improvements in the food system will reduce the costs of food products to consumers or increase their benefits.

VCs, as engines of growth, create added value that has five components:

1] salaries for workers;
2] a return on assets (profits) to entrepreneurs and asset owners;
3] tax revenues to the government;
4] a better food supply to consumers; and
5] a net impact on the environment, positive or negative.

This value added sets in motion three growth loops that relate to economic, social and environmental sustainability, and directly impacts poverty and hunger. The three growth loops are: (1) an investment loop, driven by reinvested profits and savings; (2) a multiplier loop, driven by the spending of increased worker income; and (3) a progress loop, driven by public expenditure on the societal and natural environments.

Beyond commercial and fiscal viability, the sustainability element of SFVCD involves a shift to institutional mechanisms that lead to a more equitable distribution of the increased value added and to a reduced use of and impact on non-renewable resources. The three sustainability dimensions are closely related: social and environmental sustainability increasingly become issues that determine market access and competitiveness.

Initially, SFVCD focuses mostly on efficiency improvements that reduce food prices and increase food availability and thus allow households to buy more food. However, as their incomes increase, households tend to spend more money on higher-value food (i.e. food with improved nutritional value, greater convenience, health benefits or better image) rather than increase the amount of food they consume. In turn, this evolution of consumer demand becomes a core driver for innovation and value creation at each level of the food chain, leading to continuous improvement in the food supply and increasing benefits to consumers.

This paradigm exposes a number of fallacies relating to food chain development, such as: smallholder farming should be preserved; "value chain development can only help a small minority of farmers; and the problem of food insecurity can be solved within the food system.

PRINCIPLES OF SUSTAINABLE FOOD VALUE CHAIN DEVELOPMENT

SFVCD calls for a particular approach to analysing the situation, to developing support strategies and plans, and to assessing developmental impact. This is captured in this publication by ten interrelated principles.

The approach is not about simply developing long lists of often well-known constraints and then recommending that they be tackled one by one. Rather, SFVCD takes a holistic approach that identifies the interlinked root causes of why value-chain actors do not take advantage of existing end-market opportunities.

The ten principles are grouped in three phases of a continuous development cycle.

In the first phase, measuring performance, the VC is assessed in terms of the economic, social and environmental outcomes it delivers today relative to a vision of what it could deliver in the future (Principles 1, 2, and 3). SFVCD programmes should target VCs with the greatest gap between actual and potential performance.

In the second phase, understanding performance, the core drivers of performance (or the root causes of underperformance) are exposed by taking three key aspects into account: how VC stakeholders and their activities are linked to each other and to their economic, social and natural environments (Principle 4); what drives the behaviour of individual stakeholders in their business interactions (Principle 5); and how value is determined in end markets (Principle 6).

The third phase, improving performance, follows a logical sequence of actions: developing based on the analysis conducted in phase 2, a specific and realistic vision and an associated core VC development strategy that stakeholders have agreed upon (Principle 7), and selecting the upgrading activities and multilateral partnerships that support the strategy and that can realistically achieve the scale of impact envisioned (Principle 8, 9, and 10).

CONCLUSIONS

SFVCD provides a flexible framework to effectively address many challenges facing food-system development. In practice, a misunderstanding of its fundamental nature can easily result in limited or non-sustainable impact. Even if practitioners understand and rigorously apply the principles of SFVCD, this approach cannot solve all problems in the food system. Food VCs cannot provide incomes for everyone, cannot incorporate trade-offs at the food-system level and cannot entirely avoid negative environmental impacts. Public programmes and national development strategies are needed to address these limitations. However, such programmes and strategies are largely financed through tax revenues generated in value chains, thus placing VC development in general, and SFVCD in particular, at the heart of any strategy aimed at reducing poverty and hunger in the long run.

Abstract

Using sustainable food value chain development (SFVCD) approaches to reduce poverty presents both great opportunities and daunting challenges. SFVCD requires a systems approach to identifying root problems, innovative thinking to find effective solutions and broad-based partnerships to implement programmes that have an impact at scale. In practice, however, a misunderstanding of its fundamental nature can easily result in value-chain projects having limited or non-sustainable impact. Furthermore, development practitioners around the world are learning valuable lessons from both failures and successes, but many of these are not well disseminated. This new set of handbooks aims to address these gaps by providing practical guidance on SFVCD to a target audience of policy-makers, project designers and field practitioners. This first handbook provides a solid conceptual foundation on which to build the subsequent handbooks. It (1) clearly defines the concept of a sustainable food value chain; (2) presents and discusses a development paradigm that integrates the multidimensional concepts of sustainability and value added; (3) presents, discusses and illustrates ten principles that underlie SFVCD; and (4) discusses the potential and limitations of using the value-chain concept in food-systems development. By doing so, the handbook makes a strong case for placing SFVCD at the heart of any strategy aimed at reducing poverty and hunger in the long run.

About the author

David Neven is a Marketing Economist at FAO Rural Infrastructure and Agro-Industries Division (AGS). Based in Rome, Italy, he provides technical guidance to projects and governments on the development of agrifood markets, agribusinesses and food value chains. He holds a Ph.D. in agricultural economics from Michigan State University, United States of America.

Acronyms

CSF critical success factor
GAP good agricultural practices
IBM inclusive business model
ICT information and communication technology
ILO International Labour Organization of the United Nations
KTDA Kenya Tea Development Agency
PPP public–private partnership
SFVC sustainable food value chain
SFVCD sustainable food value chain development
SMAE small and medium-sized agro-enterprises
USP unique selling proposition
VC value chain

CHAPTER 1
Introduction

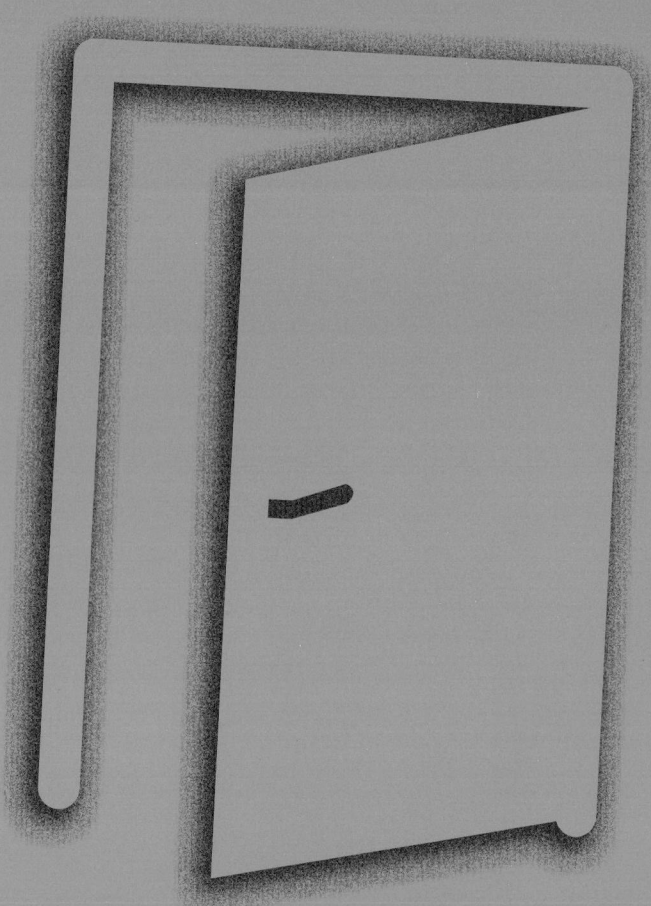

Introduction

Over the last decade, the value chain (VC) has established itself as one of the main paradigms in development thinking and practice. This has been accompanied by rapid increase in literature dedicated to all aspects of VCs. Value-chain analysis in particular received much attention, and many general and specific guides were developed.[1] Other VC publications focused on particular aspects of the approach, such as VC selection, strategy development, implementation plans and tools for analysing the enabling environment.

The sheer volume of literature on VCs, and the many variations on definitions and approaches, has made it difficult to see the big picture. Although rapidly emerging as a key theme in more recent VC publications, the triple-bottom-line approach to sustainability – combining economic, social and environmental aspects – has not yet received a thorough systematic treatment in the literature. Furthermore, much practice-based learning on VC development, including on its potential and limitations, remains restricted to small audiences. At the same time, the VC framework has largely remained in the toolkit of development practitioners and as such is not well grounded in science.[2]

Against this background, the Food and Agriculture Organization of the United Nations (FAO) has initiated a new set of handbooks on sustainable food value chain development (SFVCD),[3] and this publication is the first in the set. The handbooks aim to provide practical guidance on SFVCD by facilitating the dissemination of innovative solutions that emerged from the field to a target audience of policy-makers, project designers and field practitioners. Given the public-sector nature of the target audience, the handbooks adopt a mostly development-oriented perspective, addressing the question of how the VC approach can be used to reduce poverty and eradicate hunger at scale.

The objective of this handbook is to provide a common understanding of the SFVCD concept that will underpin the subsequent handbooks. Specifically, this first handbook aims to achieve four objectives: (1) to clearly define the concept of a sustainable food VC; (2) to present and discuss a development paradigm that integrates the multidimensional concepts of sustainability and value added; (3) to present, discuss and illustrate ten principles that underlie SFVCD; and (4) to discuss the potential and limitations of using the VC concept in food systems development.

This publication is thus not another "how to" VC analysis manual. Rather, it aims at providing a solid conceptual foundation on which to build the practical guidance that the rest of the handbooks will provide. It intends to simultaneously operationalize the insights emerging from academic discourse and to codify the lessons learned by practitioners. Drawing from best international practice, the

---

[1] See, for example, Donovan *et al.* (2013) for a comparative review and da Silva and de Souza Filho (2007) for a specific example from FAO.

[2] Gómez *et al.* (2011) present a framework for giving a better scientific grounding to assessments of the performance of food VCs.

[3] For clarity, this term refers to the development of food value chains that are sustainable.

handbooks discuss particular aspects of SFVCD, such as input supply systems, inclusive business models, producer organizations, post-harvest technology, investment promotion, territorial development, the greening of VCs and a variety of other topics.

The particular focus of this set of handbooks is on food VCs that connect farmers or fisherfolk to the final food consumers. While the general principles of sustainable VC development do not differ much across different products, food VCs do have four unique characteristics that distinguish them from other VCs in their specifics:

1] All of us are part of the food VC; we are all consumers whose well-being is directly affected by the food we eat. How food, through its nutritional value and its ability to carry pathogens, affects our health is a societal concern that necessitates rigorous supervision by the public sector. The consumer's residential location, concerns, habits and preferences related to food have a strong impact on the nature of the VC.
2] In most developing countries, agriculture and food represent a large, if not the largest, part of the economy, especially in terms of the number of people deriving an income from it. Food VCs are particularly important for the poor and impact food security directly. As such, food VCs are of strategic importance in national (and global) politics, which in turn often directly impacts the business environment in which VC actors operate.
3] Food production is closely tied to the natural environment (soils, water bodies, air, genetics) and the life cycle of plants and animals. It is thus influenced by factors that are, to a varying degree, beyond the control of producers (climate, diseases) and has social and environmental impacts that are increasingly moving from externalities[4] to internalized production costs.
4] Associated with the previous points, the quality of food products is difficult to control both in terms of uniformity (mostly at the farming stage) and in terms of preservation over time (perishability). This necessitates institutional, organizational and technological upgrading throughout the food VC (e.g. certified seed, good agricultural practices, contracts, standards, cold chains, information and communication technology [ICT]).

THE VALUE CHAIN IS A KEY CONCEPT IN THE DEVELOPMENT OF SUSTAINABLE FOOD SYSTEMS

This handbook is structured as follows.

Chapter 2 introduces the concept of a sustainable food VC, contrasts it with related but distinct concepts and places it in a sustainable VC development framework. It stresses the importance of VCs as dynamic, market-driven systems in which vertical coordination (governance) is the central dimension.

---

[4] Externalities are costs or benefits that are not transmitted through prices. They are incurred by parties who are not a buyer or seller of the goods or services that cause the cost or benefit. Thus, for example, a factory may not have to pay for cleaning its polluted waste water, but pollution it causes can represent a cost to nearby fishermen.

Introduction

Chapter 3 presents the SFVCD paradigm. Connecting FAO's overall goal of eradicating hunger to the VC concept, this presentation largely revolves around the central theme of integrating the various dimensions of sustainability and value into a single paradigm. The chapter concludes with a discussion of several development fallacies related to SFVCD.

In Chapter 4, the handbook systematically works through ten illustrated principles that underpin a continuous performance improvement in three phases of the VC development cycle:

» The first phase, **measuring performance**, assesses a VC in terms of the economic, social and environmental outcomes it delivers relative to its potential (Principles 1, 2 and 3).
» The second phase, **understanding performance**, exposes the root causes of underperformance by taking into account how VC stakeholders and their activities are linked to each other and to their economic, social and natural environment in a system (Principle 4); how these linkages drive the behaviour of individual stakeholders in terms of their commercial behaviour (Principle 5); and how value determination in end markets drives the dynamics of the system (Principle 6).
» The third phase, **improving performance**, follows a logical sequence of deriving a core VC development strategy based on the analysis conducted in phase two and the vision stakeholders have agreed on (Principle 7) and selecting upgrading activities and multilateral partnerships that can realistically achieve the scale of impact envisioned (Principles 8, 9 and 10).

Chapter 5 briefly discusses the potential and limitations of food VC development and Chapter 6 draws conclusions.

CHAPTER 2
Concept and framework

DEVELOPING SUSTAINABLE FOOD VALUE CHAINS – GUIDING PRINCIPLES

2.1 » DEFINING THE CONCEPT

There are many definitions of the VC concept in the literature. These fall into two main categories: **descriptive/structural** (what a VC is) and **normative/strategic** (how a VC should be). This publication uses a strategic definition, because this corresponds best to the central question of the practitioner: which policy/project/programme strategies should be adopted to develop a particular VC in a particular country?

Concept and framework

For the purposes of this publication, a **sustainable food** VC is defined as:

> *the full range of farms and firms and their successive coordinated value-adding activities that produce particular raw agricultural materials and transform them into particular food products that are sold to final consumers and disposed of after use, in a manner that is profitable throughout, has broad-based benefits for society and does not permanently deplete natural resources.*[5]

The "full range of farms and firms" refers to both VC actors who take direct ownership of the product and various business service providers (e.g. banks, transporters, extension agents, input dealers and processors who charge a fee). Their behaviour and performance is strongly influenced by the particular business environment in which they operate.

The term "coordinated" here means that in VCs the governance structure moves beyond a series of traditional spot-market transactions, with some level of non-adversarial vertical coordination in at least some part of the chain (following Hobbs, Cooney and Fulton 2000). This also implies that competition increasingly takes place between entire chains (or networks), rather than between individual firms. Increased coordination is part of the modernization of food VCs led by large processors and supermarket chains, but is equally important for developing VCs for staple foods that are currently traded informally.[6]

The concept of value added is central in both the definition used in this publication and the development model the author presents. Value can be added to an intermediate agrifood product not only by processing it, but also by storing it (value increasing over time) and transporting it (value increasing over space). For the VC stakeholders,[7] value added is here more formally defined as the difference between the non-labour costs incurred to produce and deliver a food product and the maximum price the consumer is willing to pay for it. Non-la-

---

[5] The definition here is mainly a variation on and expansion of the definition by Kaplinsky and Morris (2000).

[6] See, for example, Reardon *et al.* (2012) for a discussion on VC development for staple foods in Asia.

[7] Value-chain stakeholders are all those who have a stake in the performance of the value chain, including farmers, other agribusinesses, for-profit and not-for-profit service providers, consumers and the government.

bour costs are all costs other than salaries paid to casual or permanent employees. As such, the value that is created in a VC is captured in five ways:

1] salaries for employees;
2] net profits for asset owners;
3] tax revenues, including illegal forms of "taxation" associated with corruption and extortion;
4] consumer surplus, which is the difference between what the consumer is willing to pay for the product and the actual market price paid for it;
5] externalities, which represent a fifth dimension of value added. The activities taking place inside the VC will inevitably affect the wider environment, broadly defined. Externalities include negative effects (cost to society) such as air pollution caused by an economic actor but not paid for by them and positive effects (value to society) that farms and agribusinesses have on the environment but are not getting paid for, such as increased biodiversity in farming areas or effects of inputs used in one VC that spill over to another VC. The value added to society takes these broader environmental impacts into account.

Figure 1 illustrates this breakdown of the concept of value added.

THE VALUE CREATED IN FOOD VALUE CHAINS IS CAPTURED IN FIVE WAYS

FIGURE 1
A breakdown of the concept of value added

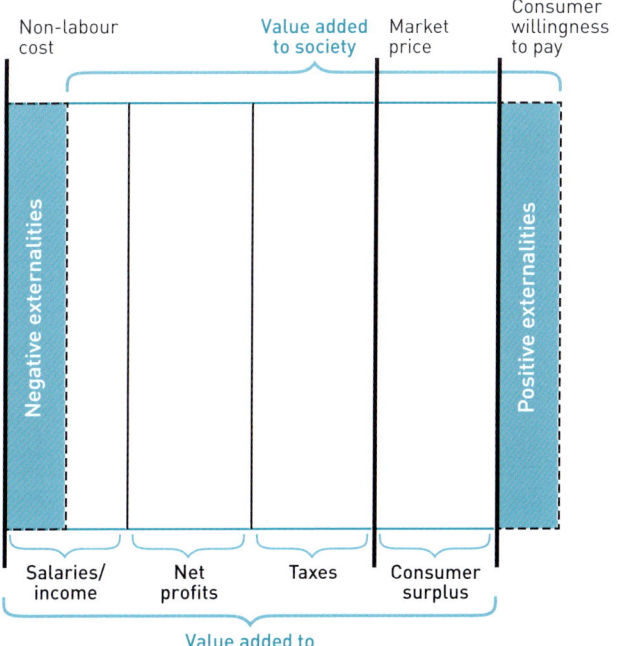

Source: author.

DEVELOPING SUSTAINABLE FOOD VALUE CHAINS – GUIDING PRINCIPLES

Concept and framework

Commercially, the main objective of VCs is to maximize profits not only by eliminating inefficiencies but also by maximizing aggregate revenues for all actors in a particular VC by creating products that consumers are willing to pay more for or buy more of. In other words, the main objective of a VC is to efficiently capture value in end markets in order to generate greater profits and create mutually acceptable outcomes for all farms and firms involved in the VC from production to consumption and disposal. Furthermore, it should be noted that value can be added or lost at each stage, e.g. post-harvest losses may occur during storage and packing.

Aspects of social impact, most notably equitable distribution of the value added along the chain and of the environmental footprint of the chain, are increasingly intertwined with the core aspect of competitiveness in the VC in at least two ways. First, trade-offs may need to be made, such as adoption of greener operations that may result in a less-competitive price. Second, social and environmental sustainability are themselves becoming sources of value creation and competitiveness. For example, a greener product image may represent a higher value to consumers and (positively) differentiate the product in the market.

VCs are meso-level structures, falling between the macro-level of the national economy and the micro-level of the individual actor. As such they can be interpreted in a narrow sense (the firms and functions leading to a particular product on the shelf; e.g. a 500 g pack of brand Z minced beef in supermarket Y in town X) or in a broad sense (all the firms and functions involved in production of a broad category of related food products; e.g. the bovine products from country Z in a range of markets where they typically compete with competitive products from other countries). This handbook, which looks at the overall developmental impact of VC growth, mostly refers to VCs in the broad sense.

2.2 » RELATED CONCEPTS

There are several concepts that are related to the VC concept, such as the *filière* (commodity chain) and the supply chain. However, although these terms are often used interchangeably, they represent distinct notions. These concepts have developed over time to address the limitations of older concepts, with newer concepts superseding older ones. Seven of these concepts – *filière*, supply chain, subsector, Porter's VC, global commodity chain, net-chain, inclusive business model, food system and landscape system – are discussed briefly here and are presented in more detail in the Annex, Concepts related to the value chain concept.

The *filière* approach (also referred to as the commodity chain approach) is the earliest of these concepts, dating from the 1950s. Initially, this approach focused on optimizing physical product flows and conversion ratios related to the large-scale processing of commodities, mostly export crops such as cocoa. Over time the concept has been broadened and today it largely coincides with the VC concept.

The 1980s were a rich period in terms of concept development, seeing the parallel emergence of the concepts of the subsector, the supply chain and Porter's VC, each of which expanded on the *filière* approach in various ways.

Driven by rapid technological developments and industrialization, **the supply-chain** concept added elements from business school economics such as finance, information, knowledge and strategic interfirm collaboration to the *filière* concept. Nevertheless, the supply-chain concept remains mostly concerned with the optimization of the flow of products and services through the chain, i.e. logistics.

The **subsector** approach added mapping of the flow of a particular raw commodity through various distinct, competing channels to a range of consumer markets, as well as introducing the notion that such subsectors are dynamic systems that change over time.

Porter's VC concept introduced "value chain" as a new term. It put forward the notion of value addition in competitive markets as the core element in the production-to-consumption chain of activities. However, Porter's VC concept deals essentially with firm-level strategy and not with broader economic development.

> THE VALUE CHAIN IS A DISTINCT CONCEPT

A turning point came with the introduction of the **global commodity chain** concept (Gereffi and Korzeniewicz 1994), from which the VC concept as it is understood in economic development today is largely derived. The global commodity chain concept combined elements from its predecessors and added the notion of chain governance, i.e. how various firms across the entire chain are coordinated (or strategically linked) in order to be more competitive and add more value. It also emphasized how this coordination is increasingly determined by large global buyers such as retailers and brand marketers. As such, the concept highlights that VCs are driven by two interrelated elements: the nature of final consumer markets and the process of globalization.

Variations on the VC concept have emerged since 2000 in response to perceived limitations of the original VC concept.

The **net-chain** concept brought horizontal linkages (networks) and the interaction between horizontal and vertical coordination, e.g. farmer groups having more market opportunities, more explicitly into the model.

The **inclusive business model** (IBM) addressed the particular challenge of integrating the poor in VCs, either as producers or as consumers. The IBM concept has the added benefit that its focus on a particular part of the VC, e.g. smallholder farmers linking directly to a processor, makes it more manageable than the broader and more complex VC concept. On the other hand, the very nature of the business-model approach, i.e. its more narrow focus, makes the challenge of how to achieve impact at scale more immediate.

The **sustainable food VC concept** presented in this publication fits in this context as well, as it more formally adds broadly defined dimensions of sustainability to the VC concept and applies it to the specific nature of food production, processing and distribution.

Looking to the future, concepts that are even more holistic will no doubt come to the forefront to address the limitations associated with the single-commodity

DEVELOPING SUSTAINABLE FOOD VALUE CHAINS – GUIDING PRINCIPLES

focus of the VC concept. Food VCs, for example, are not separated from each other: farmers typically handle multiple agricultural, livestock or fisheries products and have to make interrelated decisions on them (i.e. farming systems); and business services, infrastructure and policies are often not specific to a single commodity (e.g. finance, markets and land policy).

Concept and framework

Such broader concepts include the **food system**, which integrates all food VCs in a particular country into a single concept, and the **landscape system**, which integrates all interacting systems (economic, social and natural) in a particular geographic location into one concept.

These broader concepts allow for an assessment of the relative importance of one VC versus another, how various VCs interact with each other and with the wider environment and which changes in the enabling environment will likely have the greatest overall developmental impact.

Ultimately, however, the focus of the practitioner has to return to the VC concept being applied to particular food products. It is therefore expected that these new concepts will complement rather than replace the VC concept. The potential and limitations of the VC approach to the development of agrifood chains are discussed further in Chapter 5, *Potentials and limitations*.

2.3 » THE SUSTAINABLE FOOD VALUE CHAIN FRAMEWORK

The framework presented in Figure 2 builds on the many VC frameworks that can be found in the literature. In essence, it presents a system in which the behaviour and performance of farms and other agrifood enterprises are determined by a complex environment.

The framework is built around **the core VC**, which relates to the VC actors, i.e. those who produce or procure from the upstream level, add value to the product and then sell it on to the next level. Value-chain actors are mostly private-sector enterprises, but may include public-sector organizations such as institutional buyers (e.g. food-reserve agencies, emergency food buyers such as the World Food Programme, and the military). Actors at a given level of the chain are heterogeneous, with types of actors that are distinct in terms of size, technology, goals etc. linking through different channels to a variety of end markets.

Four core functions (links) are distinguished in the chain: **production** (e.g. farming or fishing), **aggregation**, **processing** and **distribution** (wholesale and retail). The aggregation step is especially relevant for food VCs in developing countries; efficiently aggregating and storing the small volumes of produce from widely dispersed smallholder producers is often a major challenge in these countries. The aggregation function can be taken on by producer groups, by intermediaries specialized in aggregation, by food processors or, less commonly, by food distributors (wholesalers or retailers).

A critical element of the core VC is its **governance structure**. "Governance" refers to the nature of the linkages both between actors at particular stages in the chain (horizontal linkages) and within the overall chain (vertical linkages). It refers to elements such as information exchange, price determination, standards, payment mechanisms, contracts with or without embedded services, market power, lead firms, wholesale market systems and so on.

Value-chain actors are supported by business development support providers; these do not take ownership of the product, but play an essential role in facilitating the value-creation process. Along with the VC actors, these support providers represent the **extended VC**.

Three main types of support provider can be distinguished:

1] **providers of physical inputs**, such as seeds at the production level or packaging materials at the processing level;
2] **providers of non-financial services**, such as field spraying, storage, transport, laboratory testing, management training, market research and processing;
3] **providers of financial services**. These are separated from other services because of the fundamental role played by working capital and investment capital in getting the VC on a path of sustained growth.

UNDERSTANDING VALUE CHAINS REQUIRES UNDERSTANDING THEIR COMPLEX ENVIRONMENT

FIGURE 2
The sustainable food value chain framework

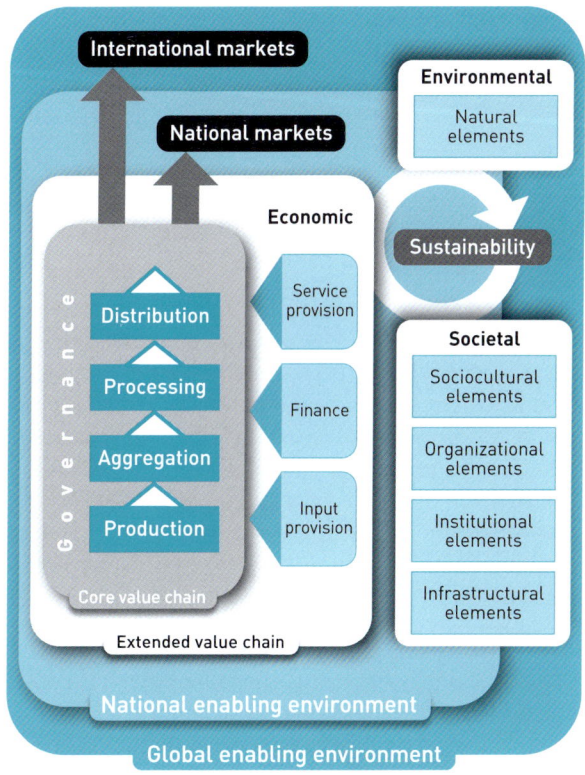

Source: author.

DEVELOPING SUSTAINABLE FOOD VALUE CHAINS – GUIDING PRINCIPLES

Concept and framework

The three types of support can in practice be delivered as a package by a single provider (e.g. seed and fertilizer, insured and on credit, with built in extension services). These support providers can be private-sector, public-sector or civil-society organizations and they can be directly part of the governance structure (e.g. services embedded in outgrower contracts).

Ultimately, value is determined by the consumer's choice of which food items to purchase on **national and international markets** (their so-called "dollar vote"). The effects of this flow down to the production, processing and support-provider levels.

Value-chain actors and support providers operate in a particular **enabling environment** in which societal and natural environmental elements can be distinguished.

Societal elements are human constructs that make up a society. These can be grouped into four types:

1] informal **sociocultural** elements, e.g. consumer preferences and religious requirements;
2] formal **institutional** elements, e.g. regulations, laws and policies;
3] **organizational** elements, e.g. national interprofessional associations and research and educational facilities;
4] **infrastructural** elements, e.g. roads, ports, communication networks and energy grids.

Natural elements include soils, air, water, biodiversity and other natural resources.

Within the enabling environment, we can further differentiate between the national environment (e.g. a country's food-safety laws) and the international environment (e.g. international food-safety standards such as the CODEX Alimentarius).

The **sustainability** of the VC plays out simultaneously along three dimensions: **economic**, **social** and **environmental**. On the economic dimension, an existing or proposed upgraded VC is considered sustainable if the required activities at the level of each actor or support provider are commercially viable (profitable for commercial services) or fiscally viable (for public services). On the social dimension, sustainability refers to socially and culturally acceptable outcomes in terms of the distribution of the benefits and costs associated with the increased value creation. On the environmental dimension, sustainability is determined largely by the ability of VC actors to show little or no negative impact on the natural environment from their value-adding activities; where possible, they should show a positive impact.

By definition, sustainability is a dynamic concept in that it is cyclical and path-dependent, i.e. the performance in one period strongly influences the performance in the next one. The sustainability concept is discussed in greater detail in Chapter 3, *The sustainable food value chain development paradigm*.

CHAPTER 3
The sustainable food value chain development paradigm

The SFVCD paradigm starts from the premise that food insecurity is a symptom of poverty. If households always have sufficient financial resources (income, wealth and support) to meet their needs, they create the effective demand that drives the supply of food.[8] On the supply side, improvements in the food system driven by competition can reduce the cost of food to the consumer or increase its nutritional value without increasing its price.

Reducing the cost of food will have a strong effect on poverty when food accounts for a large portion of household expenditure for a large part of the population, as is the case in most developing countries. Addressing hunger sustainably and in the long term thus implies addressing both an underperforming economic system and an underperforming food system. SFVCD plays a central role in this process, but needs to be accompanied by the development of sustainable non-food VCs and by programmes that improve the enabling environment, facilitate self-employment and strengthen social protection.

As illustrated in Figure 1, the value added by VCs has five components:

1] salaries for workers;
2] a return on assets (profits) to entrepreneurs and asset owners;
3] tax revenues to the government;
4] a better food supply to consumers (consumer surplus); and
5] a net impact on the environment (externalities), which may be positive or negative.

This value added sets in motion three growth loops – the investment loop, the multiplier loop and the progress loop. These loops influence economic, social and environmental sustainability and directly impact poverty and hunger (Figure 3). While infinite growth on a finite planet is not realistic, technological breakthroughs coupled with institutional strengthening will allow us to keep producing more food or higher-quality food with fewer resources for quite some time. At any rate, growth combined with equitable distribution of the associated value added is necessary to lift the poor out of poverty. The three growth loops are discussed in the context of SFVCD in the sections 3.1–3.3.

3.1 » RETURN ON ASSETS: FARMING AS A BUSINESS AND DEVELOPMENT OF SMALL AND MEDIUM-SIZED AGRO-ENTERPRISES

Net income from labour has to be in line with the value of output created by that labour, i.e. labour productivity (the value of the goods or services produced per hour worked). In order to increase labour productivity, labour needs to be paired

---

[8] FAO (2006) defines food security as having four dimensions: access (having the means to secure food), availability (food supplied in sufficient quality and quantity), utilization (healthy living through diet, sanitation, and access to clean water and health care) and stability (continuous access, availability and use). The point being made here is that, in long-term development, the driving dimension is access, in that having the money to buy food (or being entitled to food through social protection) will drive a supply response that addresses the other dimensions of food security.

3 » The sustainable food value chain development paradigm

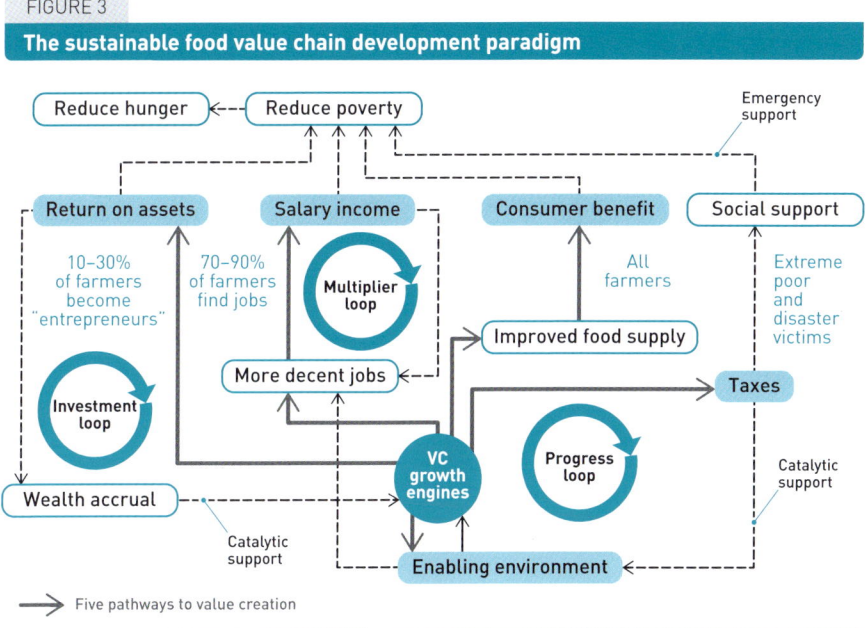

FIGURE 3
The sustainable food value chain development paradigm

Source: author.

THE SFVCD PARADIGM LINKS VALUE CHAIN DEVELOPMENT TO FOOD SECURITY

with higher levels of capitalization (e.g. agricultural mechanization), which in turn requires increased investment and working capital. This capital can be derived from retained profits or, more commonly, from lending by a growing financial sector driven by accrual of domestic wealth (investment loop in Figure 3).

Over time, increasing productivity of farm labour will commonly be accompanied by an increase in farm size, with resources, including land, shifting from less-competitive farms to more-competitive farms and a displacement of family labour by wage labour.

The shift of resources does not necessarily mean a shift in ownership (e.g. land can be rented out, providing an income to the owner). The growing farm businesses will increasingly depend on specialized enterprises both for their farming operations (e.g. input providers and land preparation services) and for marketing their output (e.g. facilitators and processors). This presents many opportunities for the development of small and medium-sized agro-enterprises (SMAE).

This shift reflects the fact that smallholder farmers are a heterogeneous group, ranging from those for whom farming is a business that they seek to expand through investment, to those who are net food buyers and subsistence farmers for whom farming is part of a survival or livelihoods strategy, i.e. a transitional strategy toward a more specialized and reliable income.

It has to be recognized that commercial farming is a form of entrepreneurship and that only a fraction of smallholder farmers (perhaps 10–30 percent) can be expected to succeed as entrepreneurs in competitive food chains.

DEVELOPING SUSTAINABLE FOOD VALUE CHAINS – GUIDING PRINCIPLES

3.2 » SALARY INCOME: CREATING DECENT WORK OPPORTUNITIES

Wages in the food VC can increase as productivity of farm labour increases and as more value is added to raw agricultural materials further downstream, but at the same time less labour will be required to produce more food (in relative terms).[9] The majority of current smallholder farmers (perhaps 70–90 percent) ultimately will have to escape poverty by securing decent work outside the farm sector.[10]

This will release farming labour, which will have to be captured by job growth elsewhere:

» with support providers;
» further downstream in the food VC, where most value is added;
» in non-food VCs; and
» self-employment.

While many of these jobs will emerge in rural areas where the commercial farms and SMAEs are located, most of these new jobs will be in urban areas (in large agribusinesses, food wholesalers and retailers and non-food industries). In both locations, but especially in rural locations, entrepreneurs in VCs and the workers they hire will spend their rising incomes on products and services, many of which will be provided by the self-employed (multiplier loop in Figure 3). As much as possible, this massive shift from agriculture to other industries should be managed as a steady, gradual process in which education (especially vocational training), mobility and urban development are indispensable elements.

3.3 » TAXES: ECOSOCIAL PROGRESS

As VCs develop, they become larger, more profitable and more formal. This increases the tax base and thus makes improvements in the enabling environment, including education and urban infrastructure, more fiscally sustainable. Since tax revenues are largely derived from value added in VCs, value-chain development also contributes significantly to financing safety nets for those who lose their source of livelihood or are hit by natural or human-induced disasters (social support).

Driven by political will and entrepreneurship, the development of the private and the public sector go hand in hand, with public–private partnerships (PPPs) providing both efficient solutions and effective coordinating mechanisms.

---

[9] This goes to the heart of the division of labour as a source of economic growth. Food is the one product we cannot live without even in the short term and therefore its supply is essential. Only if, through improvements in food productivity, one person is able to produce food for greater numbers of people can others specialize in other products, in services or in the governance of a state.

[10] The International Labour Organization (ILO) defines decent work as productive work for women and men in conditions of freedom, equity, security and human dignity. It involves work opportunities that: are productive and deliver a fair income; provide security in the workplace and social protection for workers and their families; offer prospects for personal development and encourage social integration; give people the freedom to express their concerns, to organize and to participate in decisions that affect their lives; and guarantee equal opportunities and equal treatment for all (ILO 2007).

In addition to commercial and fiscal viability, the sustainability of food VCs depends on the implementation of institutional mechanisms that lead to a more equitable distribution of the net income (or value added) arising from the VC and to a reduction in the use of and impact on non-renewable resources.

Mechanisms to achieve equitable distribution of the benefits of VCs include policies on, for example, wage labour and asset registration (e.g. land titles). Institutions to reduce impact on non-renewable resources include the introduction of environmental standards, tax incentives and markets for environmental goods (e.g. the carbon credits market).

As incomes go up, the social and environmental impacts of the food system become more important to consumers and governments, and they are subsequently increasingly incorporated in business models and food production costs (progress loop in Figure 3).

Ultimately, in a developed economy, all households should be able to derive net incomes from jobs or entrepreneurial activities that not only allow them to be food secure but also provide them with incomes that allow them to live comfortably, send their children to school, pay for housing and medical needs and handle food price spikes without going hungry.

Initially, SFVCD will mostly focus on efficiency improvements that lead to lower food prices and greater food availability and thus allow households to buy more food. However, as household incomes increase, households will tend to spend more money on higher-value food (i.e. improved nutritional value, greater convenience or image) rather than increase the amount of food they consume.

In turn, this changing consumer demand becomes a core driver for innovation and value creation at each level of the food chain, leading to continuous improvement in the food supply and increasing benefits to consumers.

> THE SFVCD PARADIGM LINKS VALUE CHAIN DEVELOPMENT TO FOOD SECURITY

3.4 » FOOD CHAIN FALLACIES

The SFVCD paradigm exposes a number of fallacies relating to development of food VCs.

Fallacy 1
Small is beautiful; urbanization is a problem; smallholder farming should be preserved

Most food in developing countries is produced by smallholder farmers (e.g. it is estimated that 90 percent of Africa's food is produced by smallholders). The inverse relationship between farm size and land productivity is overwhelmingly supported by numerous studies (e.g. Berry and Cline 1979; Cornia 1985; Carter 1984; Heltberg 1998).

Smallholder farmers use land for multiple purposes simultaneously (e.g. multiple crops and small livestock). This increases and diversifies the benefits per unit of land while simultaneously reducing pressure on natural resources relative to large-scale monocropping. Cash-poor smallholder farmers also use fewer chemicals, more natural farm inputs and more labour than large-scale commercial farmers, which helps keep their environmental footprint small.

Smallholder farmers are numerous and among the poorest of the poor; many migrate in desperation to urban areas in search of a better life, putting a strain on services and amenities. These facts seem to suggest that improving smallholder farming could contribute greatly to poverty reduction.

However, this conclusion is flawed by its lack of qualification and its blending of development and social objectives.

First, smallholders consume a large portion of the food they produced, reflecting dire economic circumstance, not economic opportunity. Even in countries where smallholders produce the bulk of the food, only a small portion of the food marketed is produced by smallholders.

Second, neither small nor large farms are always the commercially better option. Rather, there is a range of optimal farm size that depends on the nature of the crop, the natural environment and the structure of the agrifood system. Furthermore, and most importantly in terms of poverty reduction, it is not land efficiency that is the key performance measure here, but labour productivity in terms of value of output per unit of labour. Value of output is not only determined by volume, but also by the ability to sell at a good price. For smallholder producers, this ability is undermined by high transaction costs, low market power and limited access to finance, services and infrastructure. While these small-farm disadvantages can be partly overcome through collective action, there is a minimum scale of operations, which varies by commodity, below which commercial viability is unrealistic. Many smallholder farmers in developing countries today fall below this threshold size.

Third, the informality of most smallholder farming makes the enforcement of environmental standards near impossible, thus undermining the perceived greener image of small-scale farming. This green image is typically earned at the cost of lower labour productivity, and thus lower income and higher levels of poverty.

Fourth, urban areas can offer more job opportunities and greater efficiencies in the provision of public services (e.g. education, health care, utilities) than rural areas. Opportunities for poor households in rural areas to escape poverty are limited, even if farm growth and spillover effects create new and more-rewarding jobs. Thus, development efforts and poverty-reduction programmes should invest in faster and smarter urban development that creates rewarding jobs in urban areas and combine this with investments that help the rural poor to secure these jobs.[11] Where they can be competitive, agro-industries should be developed in rural areas or in or near new or existing urban centres (through

---

[11] This echoes recommendations from the 2009 and 2013 *World Development Reports* (World Bank 2009, 2013).

the development of food parks, for example) as this simultaneously creates jobs in rural and/or urban areas and increases demand for agricultural raw materials.

Ultimately, traditional smallholder farming will not be able to achieve high levels of labour productivity because it is characterized by undercapitalization and derives its competitiveness from low-cost family labour. While smallholder farmers are part of the solution in the early stages of development (and for the foreseeable future), the ultimate objective is not to assure their survival but to facilitate the transition of some of them to become sufficiently large, commercially viable farming operations and help others to transition out of agriculture altogether.

The percentage distribution among these two groups will vary by location, by stage of development and by commodity. However, trying, under some "no farmer left behind" strategy, to keep all smallholder farmers, or even worse the poorest farmers, in farming and in rural areas may actually hinder large-scale poverty reduction and thus the sustainable eradication of hunger. There is a fine line between helping smallholder farmers to survive in the short term and prolonging their misery in the long term. The objective of SFVCD is not to preserve smallholder farming, it is about broad-based job creation, income growth, and wealth accrual.

> THE SFVCD PARADIGM EXPOSES A NUMBER OF FOOD CHAIN FALLACIES

Fallacy 2
The development of food value chains can help only a small minority of farmers, therefore we need to look beyond value chain development

Two misconceptions on the nature of VC development are at the basis of this fallacy:

1] The food-VC concept does not apply only to high-value agrifood products for export markets or supermarkets that set demanding standards. Rather, it applies to any agrifood product and any market. Informal markets for staple foods, which involve large numbers of smallholder farmers, are exposed to the same environmental pressures (e.g. costs and consumer demands) as formal markets and thus will, like VCs for higher-value foods, have to derive market-based upgrading strategies.
2] The added value that is created in food VCs does not accrue to current smallholder farmers only as farmers but also, for example, as downstream entrepreneurs, job seekers, consumers and beneficiaries of tax-funded support programmes. For many smallholder farmers, especially subsistence farmers, and certainly for the landless rural poor, these other pathways out of poverty are more important and sustainable than farming their own farm.

Although the development of food VCs cannot include all or even the majority of the current (smallholder) farmers in a given country, it still represents the main long-term sustainable solution for alleviating poverty among this target group.

This is not to say that VC development can solve all issues. Complementary development programmes focusing on areas other than VCs are needed, for example to spur investment in job-generating "spillover enterprises" (e.g. consumer services catering to those with greater incomes), to assist the poorest of

DEVELOPING SUSTAINABLE FOOD VALUE CHAINS – GUIDING PRINCIPLES

the poor or to address environmental challenges. The main premise is that food VC development programmes are specifically geared toward facilitating commercially and fiscally viable improvements of the food system. The absence of such viability implies a social support strategy that is not sustainable and that is appropriate only during temporary transition phases (e.g. during a protected "infant industry" stage) or in emergency situations.

Fallacy 3
The problem of food insecurity can be solved within the food system
Given that hunger is essentially an economic problem, solving it requires that the net incomes of the poor increase. This is near impossible to achieve through farming and food processing alone. If all farmers were to significantly increase their production and market their produce, supply would likely outstrip effective demand, resulting in dramatic price decreases and food losses.

Possible (temporary) exceptions to this include products that can be readily exported or for which new and/or rapidly growing markets exist (e.g. higher-value-added products catering to a rapidly growing middle class). If farmers produce only for their own food needs and do not market their produce, they will not receive the additional income they need to fund the investments needed to increase their productivity.

The development of food VCs thus has to go hand in hand with the development of other VCs that have clearly indentified market-growth opportunities and that can create large numbers of decent jobs. Nevertheless, the development of the post-harvest section of food VCs (between harvest and consumption), if sufficiently inclusive, can in the initial stages have the broadest impact, given its direct impact on demand for raw agricultural materials and the number of households involved in farming.

3.5 » THE SUSTAINABLE FOOD VALUE CHAIN DEVELOPMENT PARADIGM: CONCLUSION

The generic development model presented in this section identifies two key challenges. The first is the need to understand the root problems, key leverage points and approaches that will have the greatest impact for a specific VC in a specific country. The second is how to combine the capacities of the public sector, the private sector and civil society into an effective partnership (a "golden triangle") that will, ultimately, put money in the pockets of the rural poor and food on their tables.

Chapter 4 presents ten core principles for addressing these challenges in SFVCD.

CHAPTER 4
Ten principles in sustainable food value chain development

DEVELOPING SUSTAINABLE FOOD VALUE CHAINS – GUIDING PRINCIPLES

The VC development paradigm presented in Chapter 3, *The sustainable food value chain development paradigm*, calls for a particular approach to analysing the existing situation on the food VC, to developing support strategies and plans and to assessing developmental impact.

This approach is not about simply developing long lists of often well-known constraints that are then recommended to be tackled one by one. Rather, the approach consists of developing a stakeholder vision for the VC, identifying and prioritizing the most relevant set of interrelated constraints, and then developing integrated upgrading strategies and practical development plans that create synergies and that can realistically realize the stakeholder vision for the VC.[12]

The analysis of the VC is guided by the holistic SFVCD framework presented in section 2.3, The sustainable food value chain framework. Measurement of VC performance before and after upgrading is based on the multidimensional concepts of value added and sustainability.

Although each food VC is unique, with particular characteristics and requiring upgrading strategies tailored to those characteristics, ten interrelated principles underpin all SFVCD efforts (Figure 4).

Ten principles in sustainable food value chain development

The first phase of SFVCD is "measuring performance." This phase assesses a VC in terms of the economic, social and environmental outcomes it actually delivers relative to an initial vision of what it could deliver in the future (Principles 1, 2 and 3). SFVCD efforts should target VCs with the greatest gap between actual and potential performance.

The second phase of SFVCD is "understanding performance." This identifies the core drivers of performance (or the root causes of underperformance) by taking into account three key aspects: how VC stakeholders and their activities are linked to each other and to their economic, social and natural environment (Principle 4); what drives the behaviour of individual stakeholders in their business interactions (Principle 5); and how value is determined in end markets (Principle 6).

The third phase of SFVCD is "improving performance." This phase follows a logical sequence of actions: developing, based on the analysis conducted in phase 2, a specific and realistic vision and an associated core VC development strategy that stakeholders agree on (Principle 7); and selecting the upgrading activities and multilateral partnerships that support the strategy and that can realistically achieve the scale of impact envisioned (Principles 8, 9 and 10).

The cycle is then repeated, starting with an assessment of the impact of the efforts to improve performance.

---

[12] For an example of this approach applied to economic policy reforms, see Hausman, Rodrik and Velasco (2005).

4 » Ten principles in sustainable food value chain development

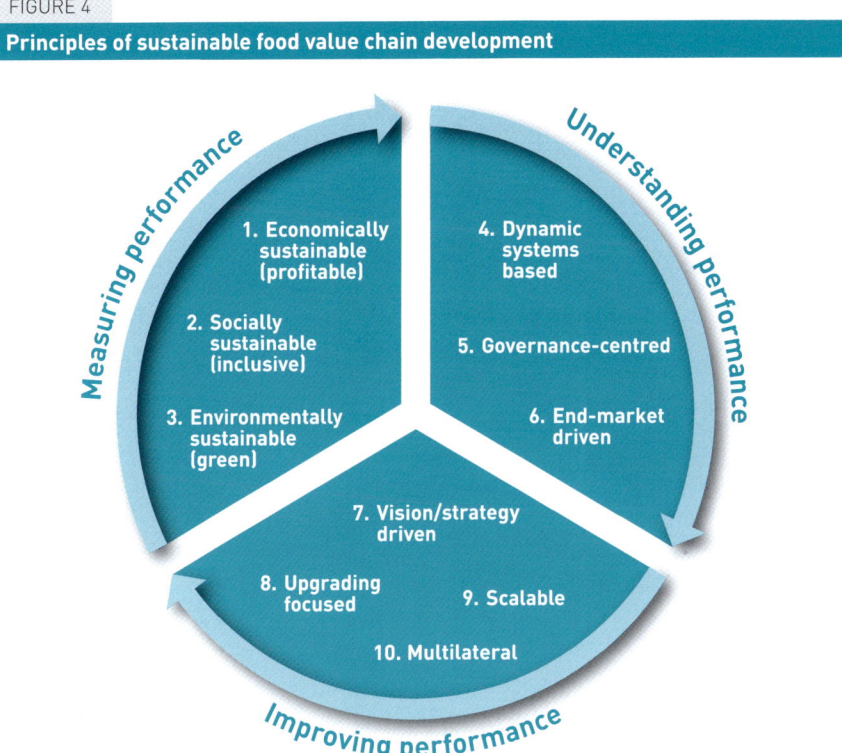

FIGURE 4
Principles of sustainable food value chain development

Phase 1: Measuring performance

Phase 2: Understanding performance

Phase 3: Improving performance

TEN INTERRELATED PRINCIPLES UNDERPIN SFVCD

Source: author.

The following sections discuss these ten principles in some detail, illustrating each of them using a case example. The cases were selected to maximize variety and address ten commodities, including examples from livestock, fisheries and crop agriculture, from ten countries across three continents. Although not all qualify fully as cases of sustainable food VCs because one or more of the sustainability dimensions may not yet have been fully addressed, each was selected because it illustrates a particular principle particularly well.

4.1 » MEASURING PERFORMANCE OF FOOD VALUE CHAINS – SUSTAINABILITY PRINCIPLES

The first three principles underpinning SFVCD relate to measuring VC performance from the perspective of the triple bottom line: economic, social and environmental sustainability.[13] These are three distinct dimensions that have a natural order in terms of timing and priority:

---

[13] The term "principles of sustainable food value chains" is not new; see, for example, Ikerd (2011).

DEVELOPING SUSTAINABLE FOOD VALUE CHAINS – GUIDING PRINCIPLES

Ten principles in sustainable food value chain development

FIGURE 5
Sustainability in food value chain development

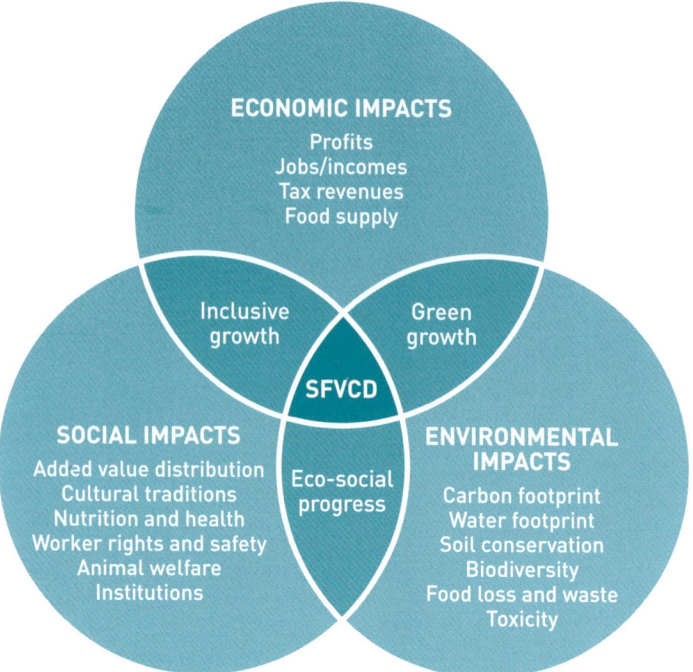

Source: author.

1] In terms of economic sustainability (competitiveness, commercial viability, growth), the upgraded VC model should provide greater (or at least not reduced) profits or incomes relative to the status quo for each stakeholder, and these should be sustained over time. Unless all stakeholders along the VC benefit, the model will not be sustainable even in the short term.

2] In terms of social sustainability (inclusiveness, equitability, social norms, social institutions and organizations), the upgraded VC model should generate additional value (additional profits and wage incomes in particular) that benefits sufficiently large numbers of poor households, is equitably distributed along the chain (in proportion to the added value created) and has no impacts that would be socially unacceptable. That is to say, every stakeholder (farmers and processors, young and old, women and men etc.) should feel they receive their fair share (win–win),[14] and there are no socially objection-

---

[14] The process toward an improved outcome may not always follow a straight line for every VC stakeholder, but go through an initial dip associated with investment and learning, before taking off.

able practices such as unhealthy work conditions, child labour, mistreatment of animals or violations of strong cultural traditions. Unless this is the case, the model will not be sustainable in the medium term.

3] In terms of environmental sustainability, the upgraded VC model should create additional value without permanently depleting natural resources (water, soil, air, flora, fauna etc.). If this is not the case, the model will not be sustainable in the long term.

Although these three sustainability dimensions are treated individually here for clarity, in practice they overlap and in some cases there will be a need for trade-offs between them (Figure 5). For example, evolving market standards, and their measurement, often have economic, social and environmental dimensions: unless all three aspects (as specified in the standard) are addressed simultaneously right from the start, the VC actors may not even be able to enter the market. In practice, some green technologies (such as perhaps conservation agriculture) may be more profitable than less environmentally friendly technologies, while others may reduce profits (e.g. the use of alternative energy sources).

Furthermore, improving social and environmental sustainability is increasingly becoming a strategic objective for agrifood firms because it determines market access (standards compliance) and may increase competitiveness (market differentiation). As such, increased social and environmental sustainability can lead to new ways to increase value creation in the food VC.

> Phase 1: Measuring performance
> Phase 2: Understanding performance
> Phase 3: Improving performance

Principle 1: SUSTAINABLE FOOD VALUE CHAIN DEVELOPMENT IS ECONOMICALLY SUSTAINABLE

PRINCIPLE 1
Sustainable food value chain development is economically sustainable

Ensuring the sustainability in food value chain development starts with the identification of sizeable opportunities to add economic value

Efforts to ensure economic sustainability focus on the added value that is created throughout the VC. This added value (additional profits, incomes, taxes and consumer surplus) has to be positive for each agent in the extended VC whose behaviour is expected to change in order to create the additional value. A possible exception to this is public-sector and civil-society organizations that participate as actors and service providers in some extended food VCs. Given their social role, these organizations may facilitate upgrading in the VC without capturing part of the value added. This can be considered sustainable if government funds are available indefinitely, i.e. they represent a recurring component of a fiscally viable annual public budget. Where public resources cannot be committed indefinitely, any upgrade that depends on public funding is clearly not sustainable and may in fact even have a negative impact as it undermines the actors' faith in opportunities for growth.

Since value is determined in the competitive setting of the end market (consumer market), value can be derived from any aspect the consumer is willing to pay for, such as better quality, flavour, brands, packaging, a particular origin or organic production.

At the same time, additional value can be derived from producing a food product more efficiently, e.g. through reduced physical losses, improved equipment and larger production volumes, and selling it at the same price as prior to the increase in efficiency. Such efficiency improvements can support the production of lower-priced food products that target more price-sensitive, poorer consumers.

Depending on the level of competition in the market, consumers will directly capture part of the added value, as market prices may well be lower than the prices consumers are willing to pay in upgraded food VCs (consumer surplus).

As noted in section 2.1, Defining the concept, VC stakeholders capture value added in four ways:

1] as increased profits by firms, or more broadly as returns to asset owners, including returns on savings and rents from leasing land;
2] as increased worker wages through more productive jobs;
3] as increased tax revenues for the government; and
4] as increased value for money for consumers buying food.

The fifth dimension of value added, i.e. the positive or negative impacts on the environment (externalities), relates mostly to social and environmental dimensions but has economic dimensions as well, e.g. income effects on households or individuals outside the VC.

Sustainability in food VCs is a dynamic concept. The generation of added value is not a one-off shift to an equilibrium at a higher level, but rather sets in motion or speeds up a process of growth and structural transformation. Increased incomes, higher product quality and lower prices fuel the demand for food products. Sustainability thus has to be assessed in a dynamic way, i.e. not only in terms of what the VC is today or at the end of a support programme but also in terms of its capacity to adapt and grow.

Increased tax revenues from a growing tax base allow governments to improve the business-enabling environment in fiscally sustainable fashion. Increased profits, if reinvested carefully, set in motion a positive feedback loop that is at the heart of economic sustainability.

If profits from investments by international firms are largely repatriated through intrafirm transfers rather than being reinvested in the country of investment, the growth cycle will spin at a far slower pace. On the other hand, setting restrictions on the repatriation of profits may divert job-generating investment elsewhere, which is why investment-promotion policies typically set no such restrictions. It is a balancing act.

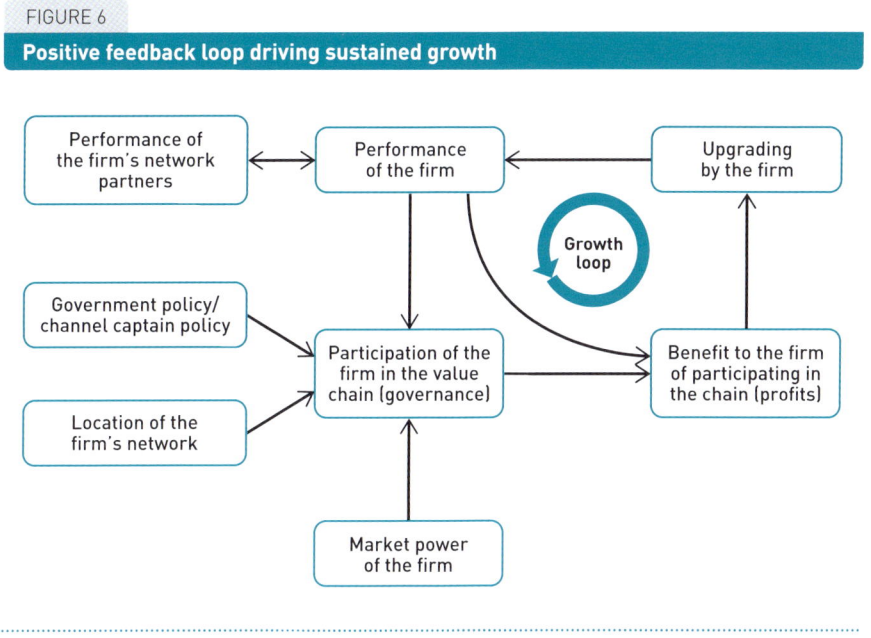

4 » Ten principles in sustainable food value chain development

FIGURE 6
Positive feedback loop driving sustained growth

```
Performance of          Performance                              Upgrading
the firm's network  ⇄   of the firm         ←                    by the firm
partners                      │                                       ↑
                              │              ↻ Growth                 │
                              ↓                loop                   │
Government policy/      Participation of the          Benefit to the firm
channel captain policy  firm in the value    →        of participating in
                        chain (governance)            the chain (profits)
Location of the               ↑
firm's network                │
                        Market power
                        of the firm
```

Source: D. Neven, 2009.

Phase 1: Measuring performance

Phase 2: Understanding performance

Phase 3: Improving performance

Principle 1: SUSTAINABLE FOOD VALUE CHAIN DEVELOPMENT IS ECONOMICALLY SUSTAINABLE

There is no such thing as a sustainable competitive advantage in business. One competitive advantage merely creates the window of opportunity to develop the next one. Adaptability to a rapidly changing business environment is the core competitive advantage.

Figure 6 illustrates how, at the firm level, economic growth can be modelled as the outcome of a positive feedback loop from performance (customer value creation) to governance structure (e.g. a contract) to profits (and other benefits) to upgrading (profit reinvestment) and back to performance.

Micro and small agribusinesses in developing countries, including commercial smallholder farmers, typically do not keep records and financial literacy levels are often low. For the greater part, these small agribusinesses have only a vague idea of their profitability. This complicates the assessment of their profitability and the sustainability of any investment in boosting their productivity.

In VC development programmes, profitability assessments are frequently either left out or executed faultily, commonly by setting the value of family labour and land costs far below market prices (often at zero cost). However, profitability is essential. It is the basic but often not fully assessed requirement for economic development: growth requires profits.[15]

---

[15] It should be noted, however, that profits may not materialize immediately, as investment costs and the time taken to learn new processes may initially have a negative effect (learning curve).

ILLUSTRATION OF PRINCIPLE 1
The potato value chain in India

In India's traditional potato value chain, typically there is no premium for quality and farmers are consequently not motivated to increase quality. When PepsiCo's Frito-Lay wanted to buy potatoes that met its strict quality requirements for the production of crisps, it faced a challenge. To meet Frito-Lay's quality requirements, farmers had to adopt a new potato cultivar ('Atlanta') that is suitable for processing into crisps, adopt new farming practices based on a different and more costly input mix and adopt new post-harvest practices, specifically in terms of handling, grading and sorting, storage and transport.

Clearly, farmers would adopt these upgrading activities only if they resulted in a commercially viable business. A study in West Bengal found that growing potatoes for Frito-Lay resulted in a 20 percent increase in costs relative to operating in the traditional potato chain, but that this was offset by higher revenues and resulted in gross margins 10–50 percent higher than those in the traditional chain, depending on yields and market prices.

Moreover, the financial incentive was supplemented by capacity-enhancing and risk-reducing elements under a contract growing scheme, a business model that PepsiCo has pioneered in India since 2001. These elements included: free technical extension services; free crop monitoring (i.e. early disease detection); guaranteed markets and prices; on-credit access to quality seed potatoes and other inputs; and weather-based insurance. The model is facilitated by vendors, local people hired by PepsiCo to act as a readily accessible liaison between the farmers and the firm.

This combination of economic incentives drove a rapid growth of the scheme, from 1 800 farmers producing 12 000 tonnes of potatoes in 2008 to 13 000 farmers producing 70 000 tonnes of potatoes in 2013. Interestingly, over time the profit motive became less important than the risk-reduction motive. Although the price of 'Atlanta' at times fell to as little as half that of the traditional cultivar, 'Jyoti' (e.g. in 2012), farmers continued to shift to 'Atlanta' because its yields are higher and more stable, and because its prices also more stable, resulting in more-reliable returns. The potatoes that fail to meet PepsiCo's quality standard (commonly 10–20 percent) can easily be sold by the farmer in the traditional market.

There are clear signals that the traditional potato channel is also modernizing, likely in part as the result of a spillover effect from the development of schemes such as PepsiCo's. This modernization includes the growth of affordable cold-storage technology (linked to extension of the electricity grid), access to price information through cell phones and adoption of improved cultivars.

Sources:
FAO (2009); Reardon *et al.* (2012); *The Hindu* Business Line (2012).

Policy and project recommendations

» Assess the profitability impact of proposed upgrading strategy for all key actors in the VC, including a financial risk analysis (sensitivity to changes in key assumptions) and make sure they are in line with the expected levels of behavioural change and poverty reduction.

» Assess if the impact of the upgrading strategy in terms of net number of jobs (salary income) created, net tax income generated, and consumer benefits provided is in line with expectations.

PRINCIPLE 2
Sustainable food value chain development is socially sustainable

The development of sustainable food value chains requires that the value added by upgrading has broad-based benefits for society and results in no socially unacceptable costs

The second dimension of SFVCD, social sustainability, refers to the critical aspect of inclusiveness. Although inclusiveness refers to equitable access to resources and markets and to having a voice in decision-making, ultimately it relates to equitable distribution of the value added relative to the investments made and risks taken. This is not only socially desirable but also amplifies the growth process through multiplier effects. The exclusion of large groups within the overall population can lead to social unrest, which undermines the sustainability of the upgraded VC.

Linked to the four economic impacts listed earlier under principle 1 (profits, incomes/jobs, food value for consumers, taxes), four dimensions of inclusiveness can be distinguished.

1] The first dimension is the number of smallholder producers and SMAEs that benefit from the upgrading strategy, i.e. that see their profits increase. Recognizing that not all smallholder producers and SMAEs can handle the upgrading proposed, the number involved should nevertheless be as great as possible, starting with the most commercial smallholders and SMAEs. Participation can be encouraged by either targeting support or by improving the enabling environment so that a process of self-selection occurs.

2] The second dimension is the number and quality of jobs that are created as a result of the upgrading strategy. These jobs include not only wage labour on farms that have upgraded but also jobs further downstream (where much of the value is added: post-harvest handling, processing, logistics etc.) and even jobs in those non-agrifood industries that benefit most from the spillover effects of increased income (e.g. local construction, small retail businesses and consumer services).

In VC development, jobs are the main path to escape poverty for the urban and especially the rural poor (e.g. subsistence farmers, landless poor).

Creating a large number of jobs and creating high-quality jobs are somewhat conflicting objectives. For example, one full-time job could replace several part-time jobs, while one job at a higher wage (based on higher labour productivity) could replace several low-paying jobs. In a normal development pattern, the number of jobs in a particular economic activity declines (at least in relative terms) while the quality of jobs improves.

3] The third dimension relates to the improved functionality of the food VC. Higher efficiency and better distribution could bring larger volumes of lower-priced food closer and on a more-reliable basis to poor consumers, including the many smallholder producers who are net buyers of food. These upgrades reduce the likelihood of price spikes for staple foods that have often led to social unrest in the past (e.g. the 2009 rice price crisis). For the higher-income segment of the market, improved standards and adding more value through, for example, processing could bring a broader variety of more convenient foods to a growing middle class. Consumers at all income levels would benefit from safer and more nutritious food products. On the consumption side therefore, food-VC development can lead to significant and broad-based benefits.

4] The fourth dimension, less direct in nature, refers to the use for social objectives of the additional tax income generated by the upgraded VC. Tax revenues can be used to fund or subsidize transition-support programmes to assist those households that are excluded from commercial food VCs or remain stuck in low-paying or part-time jobs. By focusing on capacity-building elements such as education, access to loans and information, facilitation of mobility and networking opportunities, such public programmes can facilitate a transition to more-rewarding employment opportunities. In addition, the additional tax revenues can support a social protection floor.[16]

In all four dimensions, it is not just the number of beneficiaries that matters for inclusiveness but also their distribution in terms of characteristics such as gender, income, age, location (e.g. rural or urban) and educational level. The more disadvantaged groups can benefit, the more socially acceptable and thus the more socially sustainable the outcome is.

It is also important to assess the net overall impact. For example, if certain farmers or SMAE entrepreneurs benefit from a particular programme or policy, this may come at a cost to other VC stakeholders (workers, farmers, entrepreneurs and consumers). Such costs are difficult to avoid in the context of developing

---

[16] The ILO defines the concept of a social protection floor as a social security guarantee that ensures that over the human life cycle all have basic income security and can access affordable social services in the areas of health, water and sanitation, education, food security and housing (ILO 2011).

ILLUSTRATION OF PRINCIPLE 2
The pineapple value chain in Ghana

Blue Skies Inc.'s contribution to the development of Ghana's pineapple value chain (VC) is a particularly successful example of social sustainability in food-VC development. Blue Skies, a fruit processor, was established in 1998 by a foreign direct investor with strong ties to supermarkets in Europe. Over the years the company has grown, in part by scaling up its Ghanaian activities and in part by replicating them in other countries (Brazil, Egypt and South Africa). In 2010, the firm sold 3 800 tonnes of processed fruit (pineapple and other fruits) and generated sales revenues of US$24 million from its Ghana operations. Although Blue Skies is not a social enterprise, it has promoted inclusiveness in the VC without undermining its competitiveness. Value is captured by farmers, workers, consumers and government, while negative externalities are minimized.

The supplier base of the firm is a relatively small group of around 200 commercially oriented small-scale farmers. Although not based on an outgrower scheme, and with the firm buying produce only after it has been graded at the factory gate or collection point, Blue Skies is one of the few cases in which smallholder farmers remained strongly involved after the 2004–2009 crisis in the Ghanaian pineapple VC. At that time, in order to remain competitive with South and Central American producers, Ghana switched to the new pineapple variety that markets wanted. This switch was accompanied by a shift to large plantation operations. Blue Skies continued to work with small-scale farmers, providing free training, free technical support and interest-free loans for inputs and equipment. Producers are paid without fail two weeks after delivery, at an annually agreed upon price that is higher than the cost of production, is adjusted for inflation and captures premiums associated with Fairtrade and Ethical Trade Organic certification. All financial costs for certification are carried by Blue Skies.

Blue Skies employs around 1 500 staff in its packing plant in Ghana, around 60 percent of whom are in permanent positions. In its recruitment, the processor implements what it calls a pro-diversity strategy, as a result of which 40 percent of the management team (including the general manager) are women. With salaries that are almost four times the minimum wage, a safe and healthy work environment and extensive staff amenities, these jobs easily meet the United Nation's definition of decent work.

At the same, value is captured by consumers and government. Consumers benefit from a high-quality, healthy, safe and ethically produced fresh product. Blue Skies mainly produces fresh-cut fruit pre-packed for supermarkets in Europe, where the air-freighted product arrives on the shelves within 48 hours from harvest. More recently, the processor started producing fresh juices for the local market. As Blue Skies is a formal business, it pays 32 percent tax on its net profits, which generates revenues for the Ghanaian government that help finance the running cost of, and improvements in, the enabling environment.

Finally, Blue Skies has a strong environmental component to its operations, thus reducing the societal loss of value added due to externalities. Around 50 percent of the pineapple production is certified organic. In addition, the pineapples are processed near the production area, which not only reduces the en-

> **ILLUSTRATION OF PRINCIPLE 2** *(continued)*
> **The pineapple value chain in Ghana**
>
> vironmental impact of transport but also reduces waste, because Blue Skies recycles all of its food waste as compost that is returned to farmers. Similarly, Blue Skies tracks its water and energy use per kilogram of output and constantly aims to reduce its overall environmental footprint. Blue Skies has even paid for an improvement of local roads, which benefits other kinds of economic and social activity.
>
> *Sources:*
> Webber (2007); Blue Skies (2010, 2012); GIZ (2011); Wiggins and Keats (2013).

a particular VC, but they do necessitate complementary programmes to assist those who do not benefit from other sustainable economic opportunities or transitional measures that help such people avoid a "hard landing."

Also falling under social sustainability is the need to avoid socially unacceptable outcomes beyond those associated with a possible non-equitable distribution of costs and benefits. This relates to the institutions, i.e. the "rules of the game" (business practices, policies, regulations and laws) such as, for example, those relating to the working conditions on farms and in food processing plants, the safety and nutritional value of the food and the treatment of animals during production or slaughter. They also include broader sociocultural norms and practices such as religion (e.g. halal or kosher processing or beef in India) or preferences for freshness (e.g. live fish and poultry). These norms and practices are increasingly codified in food product and process standards that determine market access and competitiveness.

> **Policy and project recommendations**
>
> » Assess to make sure that the various benefits from SFVCD, i.e. profits, jobs and food value, are equitably distributed across the VC, gender, age groups, income class and society as a whole.
>
> » Assess and minimize the likelihood of occurrence of socially unacceptable outcomes related to social institutions, cultural norms, safety and well-being.

PRINCIPLE 3
Sustainable food value chain development is environmentally sustainable

Sustainability in food chains depends on minimizing negative impacts on the non-renewable natural resources on which the agrifood system critically depends

Phase 1: Measuring performance

Phase 2: Understanding performance

Phase 3: Improving performance

Principle 3: SUSTAINABLE FOOD VALUE CHAIN DEVELOPMENT IS ENVIRONMENTALLY SUSTAINABLE

Perhaps more than any other type of VC, food VCs critically depend on and affect the natural environment, especially at the production stage. In recent times, this dependency has been highlighted by increases in climate variability and natural resource scarcity. A distinction should be made between critical environmental dependencies that threaten the survival of the VC and thus need to be tackled as soon as possible (e.g. overfishing leading to rapidly depleting stocks, or the use of banned chemicals leading to market exclusion) and environmental impacts that do not pose an immediate threat to the survival of the chain but that should ideally be addressed gradually over time (e.g. carbon emissions below the legal standard).

Reducing the risks of dependency and minimizing the environmental footprint requires greater operational control throughout the VC. This can be achieved by adopting improved practices (e.g. conservation agriculture) and through various forms of upgrading (e.g. irrigation, greenhouses, contract farming and public infrastructure).

The various elements of a food VC's environmental footprint include:

1] its carbon footprint, e.g. carbon emissions from energy used in the manufacture of fertilizer and for transport;
2] its water footprint, i.e. how much water is being used in the production and processing of food;
3] its impact on soil conservation, e.g. the depletion of nutrients and the limited availability of arable land;
4] its impact on biodiversity, e.g. loss of natural habitats and the risks associated with large-scale monoculture;
5] food waste and losses, and the associated complex links to profitability, consumer preference and packaging; and
6] release of toxins into the environment, i.e. poisonous materials released in air, soil or water bodies at any stage in the food chain.

Both the public sector and the private sector increasingly need to track their environmental impacts and demonstrate progress on this front. This has increased the importance of developing and tracking increasingly detailed environmental standards. In turn, this implies the need to develop indicators that are, in practical terms, quantifiable and meaningful. As the environmental footprint of a VC, or a particular actor in the VC, becomes more measurable, it will become increasingly feasible and mainstream practice to incorporate "greenness" as a production cost that can at the same time create value and improve competitiveness.

ILLUSTRATION OF PRINCIPLE 3
The beef value chain in Namibia

With an exported volume of around 12 000 tonnes (2010), Namibia is a relatively small player on the global beef market. As a result, it cannot compete purely on price. Its unique production landscape, where cattle production takes place in a delicate natural environment, made a differentiation strategy based on environmental sustainability a logical choice for increasing competitiveness.

Namibia's dynamic and market-driven stakeholders in the beef value chain (VC) collaborate through the Meat Board of Namibia, a public–private partnership. Through meetings, market research and technical support, the board facilitates synergies at the VC level. It is in part through this board that the Farm Assured Namibian Meat (FAN Meat) scheme was established. FAN Meat markets free-range, hormone-free beef with guaranteed animal welfare standards. It combines good agricultural practices (GAP), good transport practices, good veterinary practices and good manufacturing practices. GAP guarantees customers that at least 70 percent of the animals' diet is based on grazing. To ensure that this grazing does not destroy Namibia's fragile ecological environment, e.g. through bush encroachment, or reduce alternate economic opportunities, e.g. through loss of wildlife, the board promoted new community-based pasture-management practices and individual ranch-management practices through training and changes in the legal framework. A key element in this was that the reduced pressure on natural resources was not fundamentally based on reducing herds, but rather on better managing them (e.g. through the so-called holistic management approach, which focuses on restricted movement of the entire herd as opposed to the traditional approach of allowing the animals to roam freely). This approach both increased the amount of meat produced per hectare and reduced the environmental footprint of beef production.

The national strategy is embodied in the marketing strategy of Meatco, Namibia's largest beef processor. The firm launched its "Nature's Reserve" brand in September 2008, and shifted from selling wholesale to selling directly to high-end retailers or food-service providers. The brand allows quality-conscious consumers to distinguish Namibian beef from other supplies.

The success of this strategy is revealed by comparing the performance of the Namibian beef sector to that of its neighbour, Botswana, which has similar comparative advantages but has not adopted the same environment-based product-differentiation strategy. Namibia's exports have grown faster than Botswana's, especially in terms of volume. Namibia also exports more higher-value fresh-chilled boneless cuts, sells more into high-end markets and sells at prices that are 20–40 percent higher than those received for Botswana beef. With a larger share of the total kill sold as "quality differentiated" Namibian beef cuts, branded and packaged for retail, exporters have been able to pay their farmers premiums of US$28 million per year above the prices received by comparable South African farmers.

Sources:
van Engelen *et al*. (2012); FAO (2013a).

4 » Ten principles in sustainable food value chain development

Policy and project recommendations

» Assess in quantitative and qualitative terms to what degree the upgrading strategy reduces the environmental footprint of the food VC relative to set goals and best practice benchmarks, and adjust the strategy till these goals or benchmarks are achieved, subject to other (social and economic) goals and constraints.

Phase 1: Measuring performance

Phase 2: Understanding performance

Phase 3: Improving performance

4.2 » UNDERSTANDING FOOD VALUE CHAIN PERFORMANCE – ANALYTICAL PRINCIPLES

Unlike many other development approaches, VC development takes a holistic perspective that allows the identification of the interlinked root causes of why end-market opportunities are not being taken advantage of. The identification of these root causes essentially implies a particularly broad and dynamic interpretation of the structure–conduct–performance (SCP) paradigm (Bain 1956). This paradigm calls for an in-depth understanding of the structure of the system, of how this structure influences the conduct of the various stakeholders and of how this results in an overall performance that changes the system's structure over time.

Principle 4: SUSTAINABLE FOOD VALUE CHAIN DEVELOPMENT IS A DYNAMIC, SYSTEMS-BASED PROCESS

Principles 4, 5 and 6 underpin the analytical stage of food-VC development.

PRINCIPLE 4
Sustainable food value chain development is a dynamic, systems-based process

Only by identifying and addressing the root causes of underperformance in the system can truly sustainable food value chains be realized at scale

Value chain development starts from the premise that a VC is a system in which everything –every activity, every actor – is directly or indirectly linked. VC mapping is typically an essential part of the analysis of VC performance because we must understand the VC holistically in order to understand its performance. The VC does not operate in isolation; it is actually a subsystem that is linked to other subsystems in an overall system. An agrifood VC is linked to and influenced by market systems, the political system, the natural environment, farming systems, infrastructural systems, legal and regulatory systems, the financial system, global trade systems, social systems and many other subsystems.

As a consequence, the greatest opportunities for improving the performance of a particular VC (i.e. addressing the root causes of core problems, the real reason why something that appears to be a good idea is not happening) may lie in one of these linked subsystems rather than in the chain itself. This interdependence can take on intricate forms with cause–effect relations not always being straightforward.

Several observations follow from this.

» First, in order to achieve impact at a certain point in the system, it may be more effective to facilitate change at another entry point rather than directly at the point where the impact is to be generated. For example, to increase the market participation of farmers, it may be better to work with a bank to provide finance or with a processor to set up contract growing than to work directly with the farmers.

» Second, addressing an issue at a certain point may not have any effect on the overall system if issues at other entry points are not addressed at the same time. For example, training farmers in the use of a new piece of equipment will not result in change if farmers have no access to working capital and repair services. In other words, there is a need for integrated solutions, not solutions to individual problems.

This point is illustrated in Figure 7, which represents the flow of product (e.g. raw farm products to finished food). In Figure 7(A), resolving bottleneck 1 has little or no effect unless bottleneck 2 is addressed at the same time. For example, boosting farm productivity by providing higher-quality inputs will have little impact if high transaction costs or low product quality hinder the marketing of the increased volume.[17] In fact, the impact may be negative: the increased volumes may lead to the collapse of local market prices, which may benefit rural consumers in the short term but discourages farmers from becoming more commercially oriented in the medium term (Barrett 2008).

» Third, VC development focuses on those constraints that would have the greatest impact if resolved. Typically, these are points of leverage or binding constraints in the system where the impact of a change is greatest. The associated implication is that constraints must be tackled in the order in which they become binding, i.e. a logical sequencing of activities is critical (Demont and Rizzotto 2012). Thus, in Figure 7(B) point 3 is the leverage point as it constrains the greater "flow" channel, whereas in Figure 7(A) point 2 is the binding constraint as it has the greatest impact on "flow" through the VC. For example, it is costly and difficult to assist smallholder farmers and SMAEs individually, but many small-scale actors in the VC can be reached simultaneously through leverage points such as policies, service providers, market places and associations. In Figure 7(C), point 6, an absent link, is the leverage point because it cuts the VC off from a larger market. For example, linking smallholder farmers to new, more remote urban markets that have greater absorptive capacity and better prices may results in greater development opportunities than linking them to smaller local rural markets, even if the constraint that needs to be addressed is more challenging.

---

[17] Demont (2013), for example, shows how the impact of national rice development strategies is undermined by not investing sufficient resources in postproduction links and stages (value-adding and marketing).

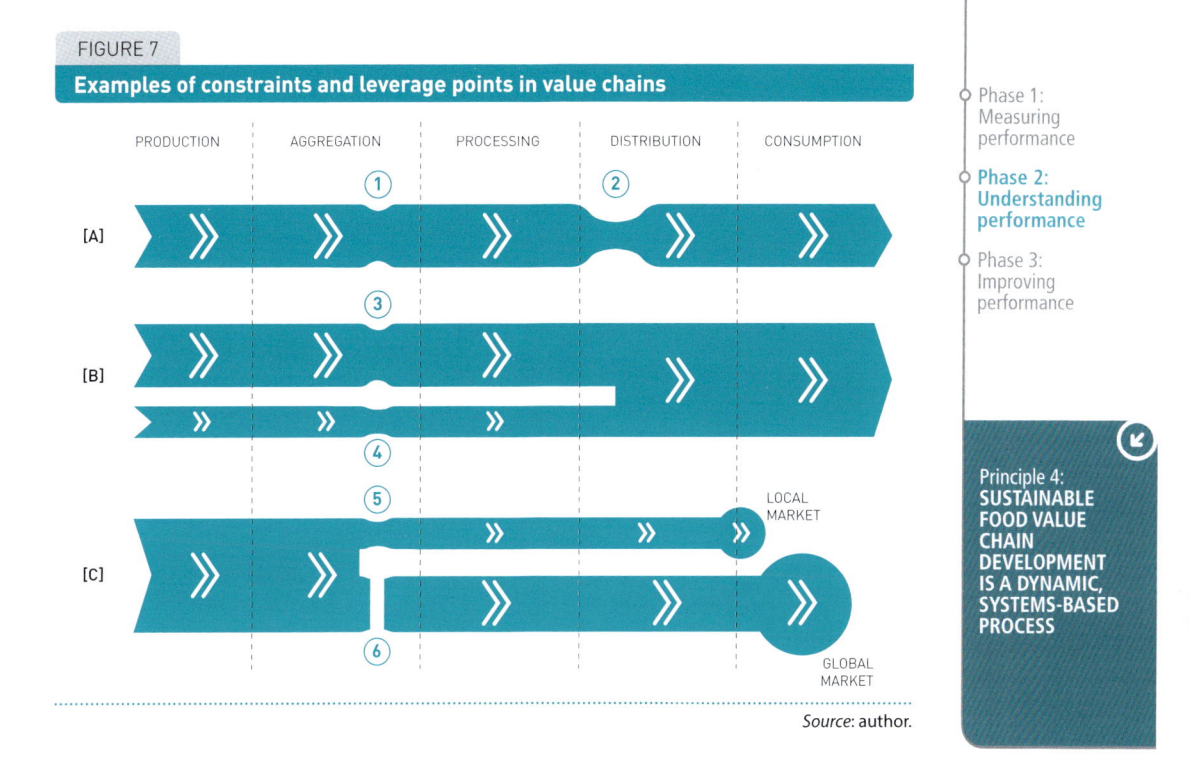

4 » Ten principles in sustainable food value chain development

FIGURE 7
Examples of constraints and leverage points in value chains

PRODUCTION | AGGREGATION | PROCESSING | DISTRIBUTION | CONSUMPTION

[A] ① ②
[B] ③ ④
[C] ⑤ → LOCAL MARKET
⑥ → GLOBAL MARKET

Source: author.

Phase 1: Measuring performance
Phase 2: Understanding performance
Phase 3: Improving performance

Principle 4: SUSTAINABLE FOOD VALUE CHAIN DEVELOPMENT IS A DYNAMIC, SYSTEMS-BASED PROCESS

The key point is that by starting from an understanding of the system as a whole, more-effective and more-efficient support strategies can be developed.

The VC is a dynamic system and it is essential to understand its dynamics (how the system evolves over time) and the factors that drive and (can) influence them. There are positive and negative feedback loops that push the system in particular directions. These can be desirable (e.g. cluster growth) or undesirable (e.g. eroding competitiveness). It is possible to influence or even reverse some of them (e.g. through government policies), while others have to be largely accepted as waves that need to be ridden (e.g. changing consumer behaviour).

The primary factors influencing the dynamics of the VC include changes in market demand, technology, available services, profitability, risk, barriers to entry, large-firm behaviour, input supply and policy. The dynamic nature of VCs and the environment in which they operate requires that VC development projects, programmes or policies must be designed to be flexible and, like the VC actors they support, able to adapt to changing circumstances.

Adaptability is the ultimate core competence for achieving great performance in the VC. Furthermore, as change does not stop when a project stops, a case can be made to take a continuous partnership approach rather than a fixed-duration project or programme approach in addressing underperformance in a the VC.

ILLUSTRATION OF PRINCIPLE 4
The vegetables value chain in the Philippines

This case illustrates, at the level of a more narrowly defined value chain (VC), how taking a dynamic systems perspective allowed the stakeholders to find the most critical bottlenecks and leverage points at each successive stage of the development of the VC.

As in many other countries, rapidly expanding supermarkets have been a key driver of change in vegetable value chains in the Philippines. The Northern Mindanao Vegetable Producers Association, or NorMinVeggies, is a new type of market facilitator that functioned as a leverage point for sustainably linking smallholder growers to these new retailers and other demanding markets. With the assistance of the United States Agency for International Development (USAID) and FAO, it has done so by identifying and addressing sets of critical constraints as they emerged.

1] Aggregation, capital and knowledge constraints: NorMinVeggies was set up in 1999 by a group of determined famers. The unique feature of the association is that it consists of two distinct but well-integrated types of farm: tiny family-operated farms with little capital investment and (mostly still) small-scale farms operated by independent part-time growers with some access to capital and technology. Combining both types of farm in crop-based marketing clusters for at least 12 different vegetables allowed family farmers to learn from independent farmers and the latter to benefit from increased aggregated volumes.

2] Quality constraint: Over the years, to meet the increasingly demanding requirements of buyers, NorMinVeggies introduced quality assurance schemes, production schedules and traceability systems. These are rigorously followed by all members, with designated lead farmers act as coaches and quality managers. The system is transparent and the responsibility for delivering quality and the benefits derived from this are shared equally among members. Individually, small family farms would not have been able to meet market requirements and post-harvest losses would have been far greater (up to 25 percent).

3] Logistics constraints: In 2006, NorMinVeggies established a consolidation centre to improve its efficiency. This centre created a leverage point not just for marketing but also for the procurement of inputs and services. The same year also saw a shift from bags to plastic crates for handling the produce, forcing other traders to follow suit. The cost of the overall system, i.e. the operational and managerial cost for the delivery of these various services to its members, is entirely covered by the value-based fees (of 2–5 percent) that members are charged, thus making the model commercially viable.

4] Market constraints: To avoid market dependencies, NorMinVeggies leveraged its larger volumes and reliable quality to bypass various layers of traditional middlemen and engage directly with a range of markets, including supermarkets, hotels, fast food chains and export, as well as traditional local and wholesale markets. Each of these markets has different requirements, necessitating constant adaptation to a changing market environment, but also allowing NorMinVeggies to sell a range of quality grades to a range of markets.

ILLUSTRATION OF PRINCIPLE 4 (continued)
The vegetables value chain in the Philippines

Over time, NorMinVeggies expanded its membership, output and range of markets. Its membership gradually increased from 15 in 1999 to 178 in 2011 and now includes individual farmers, cooperatives, foundations and growers' associations. Overall, a total of some 5 000 farmers are involved in the scheme. The system's efficiency allowed for both higher farmgate prices and lower retail prices, thus creating additional net income for farmers and increased benefits for consumers.

Sources:
Concepcion, Digal and Uy (2007); Sun Star (2011a, 2011b); author's interview with Michael Ignacio, Executive Director of NorMinVeggies (2007).

Principle 5: SUSTAINABLE FOOD VALUE CHAIN DEVELOPMENT IS CENTRED ON GOVERNANCE

Policy and project recommendations

» Invest in high-quality VC studies that use primary data to identify the root causes of observed underperformance. These studies should not be rushed and should employ skilled and experienced analysts.

» Map the VC, indicating the main channels, types of actors, leverage points and product flows in sufficient detail for strategic decision-making, while avoiding distracting overcomplication.

» Identify the dynamics of the VC system and discuss their strategic implications.

PRINCIPLE 5
Sustainable food value chain development is centred on governance

Strategies that take behavioural assumptions and governance mechanisms, and the factors that influence them, into account are more likely to result in high levels of impact

Achieving impact by improving the performance of a VC requires behavioural change by the VC actors. Specific economic behaviour results from a specific set of interconnected economic, social and environmental elements. These causal relationships must be understood in sufficient detail and threshold values for binding constraints must be exceeded if a VC development programme is to change VC actors' behaviour. For example, training will not result in behavioural change (and thus will have no impact) if the topic of the training is not the (only) binding constraint or if the training is inadequate in terms of content or delivery.

To put this another way, how actors in a VC behave (operate internally and interact externally) depends on their incentives (prices of inputs and outputs, elements of risk, culture, personal preferences, attitudes and transaction costs) and their capacities (financial, human, physical, social, informational, etc.). These incentives and capacities will differ for different types of actor (e.g. small farms versus larger farms, agribusinesses or food distributors; women versus men, old versus young, rural versus urban; and so forth). Therefore, VC development programmes must recognize the heterogeneity of actors involved in the VC.

For example, farmers will not adopt a new technology, even if it improves technical performance, if the risks are too high or its effect on profit is too small. To illustrate further, increased fertilizer use may increase yields under normal conditions but exposes a smallholder farmer to a greater risk. They will lose their scarce cash if crops fail or if the increased harvest does increase revenue sufficiently to offset the increased costs. In this situation, purchasing fertilizer would represent an economically irrational decision on the part of the farmer.

Furthermore, recognizing insights from the farming systems literature, farmers have to decide how to allocate scarce resources to a variety of farming and non-farming expenditures and this must be taken into account in developing the VC for a particular commodity. For example, a farmer may decide that payment of school fees should take precedence over purchasing fertilizer. Thus, overcoming all the root causes of low fertilizer use may require a combination of input insurance, extension advice on efficient fertilizer use (e.g. in conservation farming), improved market linkages (e.g. contracts) and even loans for school fees.

In a VC context, incentives and capacities are in large part determined by the nature of the vertical and horizontal linkages between the actors. Actors can be linked to each other vertically through governance mechanisms that range along a continuum from pure spot-market transactions, over contract mechanisms and partnerships to vertical integration, where multiple links in the VC fall within the boundaries of a single firm.

As the governance mechanism is a key determinant of how much benefit a VC actor extracts from a transaction, it is essential that it represents win–win solutions for all parties in the transaction, thus aligning incentives along the VC and facilitating behavioural change.

The nature of the governance mechanism is linked to the structure of the VC, with firms at various links in the chain varying in terms of their size, financial strength, network connections and access to information. In other words, actors differ in terms of market power. Greater market power is typically associated with firms that have the most influence over the VC (lead firms and channel captains); these are often essential partners in any development strategy.

Collective action (or horizontal coordination) by smaller VC actors (e.g. smallholder farmers or SMAEs) can both reduce differences in market power among VC actors and lower transaction costs. The nature of business service providers (e.g. inputs, finance, information and transport) and the broader enabling environment

ILLUSTRATION OF PRINCIPLE 5
The tea value chain in Kenya

The tea value chain (VC) in Kenya is one of the most successful cases of smallholder farmer inclusion in a VC, both in terms of the number of farmers involved and the degree to which they are included. This success is largely due to the unique governance structure of the chain.

Climatic conditions make Kenya well-suited for tea production and its black teas are a high-quality ingredient in the large-volume segment of the tea market. About 60 percent of all tea grown in Kenya is produced by smallholder farmers. Although smallholder yields are lower than estate yields (mainly because of less intensive effort on the part of smallholders), small-scale tea producers generate a good income from their tea because of its high quality and, mainly, because they capture a larger part of the value added further downstream.

Smallholder tea growers bring their tea to buying centres, from where it is transported to one of the 63 tea factories in the country. Each factory has about 60 buying centres (functioning as quality control points). Each buying centre has five committee members elected from the farmers who deliver tea to that buying centre. Six members of the factory board are elected from among the buying-centre committee members, with the factory board supervising a professional factory management team. The buying centre keeps a record of the exact number of plants each farmer has and thus knows how much tea each farmer is expected to deliver. Each tea factory is supplied by an average of 7 000 smallholder growers, each of whom has between 0.5 and 3 acres of tea. Each tea factory is a separate company that is fully owned by some but not all of the farmers that supply it. This is the result of a farsighted privatization scheme. There are about 450 000 smallholder tea growers in Kenya, 150 000 of whom are the exclusive shareholders of the factories.

The 63 factories in turn own the Kenya Tea Development Agency (KTDA). All smallholder producers, including the 300 000 who do not own shares in the factories, are required by law to sell through KTDA. KTDA has been a private company since 2000. It provides inputs to farmers, provides human resource services (management and secretarial staff) for the factories and markets the tea. Most of the tea is sold at auction in Mombasa, but increasingly it is sold directly to tea packers, including a growing contingent of Kenyan firms. One of these Kenyan firms, Kenya Tea Packers (KETEPA), has KTDA as its majority shareholder, so farmers even capture part of the value added at the packing level.

Since most of the profits made from the sale of smallholder tea flows back to the smallholder tea growers, Kenyan tea farmers make far more money from their tea than do their counterparts in neighbouring countries. For example, not only are factory-gate prices for "made tea" (the processed tea in bulk) 10–40 percent lower in Rwanda, Uganda, and Tanzania than in Kenya, tea growers in those countries capture around only 25 percent of this price, whereas Kenyan tea growers capture 75 percent.

The governance system is not flawless. Although farmers are the legal owners of the KTDA factories, decision-making at the factories is largely in the hands of KTDA management. Rather than receive a regular shareholder dividend, smallholders are paid a fixed amount at the time of delivery to the buying centre and receive a bonus after the tea has been sold and the processing/mar-

> **ILLUSTRATION OF PRINCIPLE 5** *(continued)*
> **The tea value chain in Kenya**
>
> keting costs and KTDA management fees have been deducted. Dissatisfaction (or impatience) with this payment structure, in part resulting from non-transparent communication between KTDA and the farmers, has resulted in farmers selling their green leaves not to KTDA but via hawkers to private companies that pay them more at the time of delivery. It also resulted in the emergence of the Kenya Union of Small-scale Tea Owners (KUSTO). Furthermore, KTDA's business strategy, which is based on heavy reliance on a limited set of buyers and little value addition (through their own packing), also exposes the overall structure to considerable market risk.
>
> *Sources:*
> CPDA (2008); Knopp and Foster (2010); FAO (2013b); KTDA website (http://www.ktdateas.com).

(e.g. policies, programmes and public infrastructure) greatly influence the vertical and horizontal interactions between actors in the VC. Thus, SFVCD efforts may need to change also the behaviour of business service providers and public officials.

Trust, in the behaviour of other VC actors and in the effectiveness of the enabling environment, is an overarching, precious asset driving the performance of the VC. Lack of trust will hinder the performance of the VC. Corruption and extortion, which drain off some of the value added in the VC, undermine the emergence of trust. On the other hand, VC-wide collaboration in pre-competitive space[18] (e.g. through industry associations) can be instrumental in building trust throughout the VC.

> **Policy and project recommendations**
>
> » Analyse in detail how value chain actors of different typology transact vertically and how they collaborate horizontally.
>
> » Identify the root causes for observed behaviour in terms of how farmers and agribusiness entrepreneurs operate their businesses and how they link to their suppliers and buyers (i.e. keep asking and answering the "why" questions).

---

[18] Pre-competitive space refers to an area where public and private stakeholders collaborate without affecting their competitive position *vis-à-vis* each other. For example, various competing firms and public-sector organizations may collaborate on a research and development project or on the development of a national product image that all parties involved benefit from.

4 » Ten principles in sustainable food value chain development

PRINCIPLE 6
Sustainable food value chain development is driven by the end market

Value is ultimately determined in the end market, and therefore any upgrading strategy has to be directly and clearly linked to end-market opportunities

Phase 1: Measuring performance

Phase 2: Understanding performance

Phase 3: Improving performance

Whether involving multinationals and global markets or SMAEs and local markets, the performance of the VC is ultimately determined by its performance in an end market, where the value of the food item is determined by the consumer's purchase decision. Given that the VC should be geared toward specific end-market opportunities, the identification and quantification of such opportunities is the starting point for every successful strategy aimed at improving the performance of a VC.

Consumers will base their decision on whether or not to purchase a product on the intrinsic qualities (e.g. physical appearance, nutritional value, taste, convenience, brand, image, packaging and country-of-origin) and the price of the product. Increasingly, however, consumers will also base their purchase decision on the process by which the food item is produced and delivered at the point of final purchase. Thus, considerations related to the environmental footprint and social impacts (negative or positive) enter into the consumer's decision process.

End markets for food are not homogenous. Different consumers have different preferences. There are price- and quality-driven segments in the mass market, and various segments in numerous niche markets. There are segments in local, national, regional and global markets. There are segments in food retail and food services (restaurant) markets, most notably modern (supermarkets) and traditional segments. Size, growth, prices, competitiveness and critical success factors (CSFs) vary widely across these various end-market segments. How a VC's (potential) strengths and (difficult to address) weaknesses align with the various segments' key success factors and how they measure up against competing offers (benchmarking) will point to its most promising opportunities.

By their very nature, markets are competitive environments that induce a Darwinian process in which, in the absence of market distortions (e.g. protective policies), only the "fittest" farms and firms survive. In this context, it is important to realize that all markets are global, in that competition with food products from other countries is only as far away as the (often shrinking) cost of bringing them into the domestic market, even though protective tariff and non-tariff barriers may keep that cost artificially high.

In the modern VCs that increasingly dominate the food system, it is typically the large processors and retailers that translate consumer demand into specific requirements for suppliers. These requirements are increasingly captured in the ever-changing and ever-more-demanding product and process standards embedded in supply contracts and often associated with traceability requirements.

However, although meeting these standards is necessary to gain access to a given market it does not guarantee successful market entry. A VC will be able to enter the market successfully only if it has a unique selling proposition (USP). Moreover, it will grow its market share and/or revenues only if it continuously improves its USP. The USP, captured and signalled through branding, can be associated with the uniqueness of the product (e.g. geographic denominations), its price, its high intrinsic quality, its high extrinsic quality (image), its availability (seasonality or volume) and so on, or any combination of such characteristics. As the USP is the outcome of all the activities along the VC chain-wide collaboration is a critical factor in achieving competitiveness.

Finally, segments grow or shrink in long-term trends or sudden shifts, and the CSFs in them change over time. For example:

» Changes in lifestyles caused by urbanization, income growth and technological change, for example, result in changes in consumer preferences (typically involving higher value and more convenient food products).
» Changes in market and trade policy (national, regional, global trade policies and agreements) can greatly alter market opportunities.
» Natural shocks (e.g. droughts, floods and outbreaks of animal or plant diseases) can suddenly change market conditions, e.g. by removing a key competitor from the market or increasing demand for a substitute product.
» Changes in institutional food procurement and distribution (food purchased and distributed through national or international organizations) can create both threats and opportunities in the market.
» Changes in storage, transport or processing technology (e.g. processing cassava for beer), changes in standards or the adoption of a particular standard by a large player (e.g. a large retailer stocking only organic or Fairtrade produce in a particular product category) can cause sudden changes in food market opportunities.

VC actors must thus target several market segments simultaneously (i.e. have a sufficiently broad market portfolio) to reduce dependency risks. They must also constantly track the evolution of the market so as to be ready to adapt to changes, leave markets that are no longer of interest or enter new or emerging markets to sustain VC performance.

Policy and project recommendations

» Identify and quantify specific market opportunities that can be realistically taken advantage of, ideally based on committed demand by particular processors or distributors.

» Indentify the critical success factors that underpin competitiveness in the identified target market segments as well as the relevant competitive advantages of the value chain to be upgraded.

ILLUSTRATION OF PRINCIPLE 6
The rice value chain in Senegal

Rice is Senegal's staple food. The country has a high potential for rice production in the Senegal River valley but imports 60 percent of the 1 million tonnes of rice consumed each year. There is thus a strong potential to develop a domestic rice value chain (VC) that can take advantage of the clear end-market opportunity. While many Senegalese consumers consider local rice from the Senegal River valley to be inferior in quality to imported rice, they also associate brands with quality, as they are used to branded rice of superior quality imported from Asia. Furthermore, recent market studies revealed that urban consumers were willing to pay a premium of 17 percent for their preferred rice brands. Given that 45 percent of buyers choose rice based on the bag it comes in rather than on a close visual and sensory examination of the grain quality, branding and brand recognition play a very important part in any marketing effort.

The main challenges in creating a competitive local rice VC are thus first to improve the quality of the rice produced, second to aggregate production and third to implement a well-designed marketing strategy. Critically, this has to be achieved on a solid, commercially viable footing. Several initiatives have been tried, with varying success. One such initiative that shows early promise is Terral rice.

Terral is a new rice brand, owned by Durabilis, a Belgium-based impact investor, i.e. a company that invests in and manages businesses with the aim of stimulating sustainable development in low-income countries. In 2006, Durabilis started by producing drinking water, for which it established a distribution system. This distribution system was subsequently used for distributing small retail packs of rice under the Terral brand. The company is engaged throughout the VC, in the production, processing, distribution and marketing of rice to Dakar and other urban markets throughout West Africa. As such, it tackles the challenges of quality, volume, financing and marketing directly.

The initial results are promising, with both suppliers and buyers eager to engage with Durabilis. In 2011, Durabilis conducted a small trial by buying 200 tonnes of rice from a traditional market, subcontracting a miller to process it and then packaging and marketing the rice to lower-income segments in the Dakar market through its own facility. The milling involved installing a small new mill, the most advanced in the region, which delivered cleaned rice equal in quality to imported rice. In terms of branding, Durabilis chose a combination of a local brand name (Terral means "welcome" in the local language) with an international symbol inspired by India. This hybrid strategy allowed Durabilis to target both the market segment that is sensitive to rice brands that are perceived as being "local" and market segments that are sensitive to rice brands that are perceived as being "foreign". The trial generated US$100 000 in revenue. In 2012, Durabilis followed up the trial by setting up a contract-farming scheme with 450 rice growers in 15 groups for two production cycles. It processed the paddy rice through a subcontracted miller and marketed 626 tonnes of white rice. This resulted in sales of US$360 000. At the same time, 25 new jobs were created in operating and managing Durabilis' rice operations. Farmers are paid higher prices than in the traditional spot market and are paid cash on delivery. Durabilis secured a loan from two non-profit social invest-

---

Phase 1: Measuring performance

Phase 2: Understanding performance

Phase 3: Improving performance

Principle 6: SUSTAINABLE FOOD VALUE CHAIN DEVELOPMENT IS DRIVEN BY THE END MARKET

> **ILLUSTRATION OF PRINCIPLE 6** *(continued)*
> **The rice value chain in Senegal**
>
> ment funds (Root Capital and Alterfin) to cover the interval between paying farmers and receiving payment from buyers.
>
> Whether or not this model will be sustainable in the long run remains to be determined. The main reason why this model might succeed where others have failed is that the driver behind the VC development is an integral part of the VC, not a temporary facilitator. Furthermore, realizing that economies of scale are crucial in the low-margin rice market, Durabilis has ambitious plans and a long-term development horizon. Plans to strengthen the supply base through a corporate nucleus farm, to develop higher-value-added products such as fortified rice, and to set up its own milling facility all have the potential to increase commercial viability and reduce the exposure to production and market risk.
>
> *Sources:*
> USAID (2009); Demont and Rizzotto (2012); Costello, Demont and Ndour (2013); Durabilis website (http://www.durabilis.eu).

4.3 » IMPROVING FOOD VALUE CHAIN PERFORMANCE – DESIGN PRINCIPLES

The first six principles describe VC performance largely in general terms. The next four principles guide the process by which a clear and detailed understanding of the current performance of the food chain can be translated into effective and efficient programmes that support or facilitate VC development. This process is arranged in three phases:

1] setting clear goals (vision) and developing an approach to achieving the goal (core competitiveness strategy);
2] developing an action plan for technical, institutional and/or organizational upgrading of the VC that can achieve results at scale; and
3] designing and implementing a monitoring and evaluation system that continuously tracks performance against the vision and that allows for adaptations where and when necessary.

PRINCIPLE 7
Sustainable food value chain development is driven by vision and strategy

Only by carefully targeting realistic development goals and targeting particular points and stakeholders in the value chain can SFVCD be effective

A core strategy links analysis to implementation in SFVCD. A core strategy indicates the main strategic thrust, i.e. a compelling theme that knits together otherwise independent activities and focuses the energies of the various stakeholders on the complementary strategic actions needed to realize a shared vision. In practice, complexity can hinder success. It is therefore important that the chosen strategy and associated development plans are kept as simple as possible even though the analytical stage will have highlighted the complexity of the VC and its environment. Careful targeting of the strategy is essential to achieve this simplicity.

The strategy must be targeted in three ways.

1] First, the strategy has to be built around a vision. This vision describes the objectives of the VC development strategy and should be realistic, quantified as much as possible and acceptable, even inspiring, to stakeholders. The vision must encompass the triple bottom line of economic, social and environmental objectives (likely reflecting trade-offs), align with national development plans and other ongoing support activities and be realistic, based on an in-depth understanding of the VC system and the resources available for support programmes.

Delivering the vision will require broad-based buy-in by stakeholders, political will and entrepreneurial drive. Individual efforts and commitment from political, community and business leaders often make the difference between success and failure in VC development. SFVCD programmes must recognize the realities of both the politics of government and the market power of large firms in the agrifood subsector, and align these realities with the VC vision.

2] Second, the strategy has to be targeted at the right stakeholders. The ultimate objective, from the public-sector perspective, is to address poverty and eliminate the associated problem of hunger, not just temporarily but sustainably. This implies that VC development has to be inclusive of the poor. Inclusive, however, does not imply a "no farmer left behind" strategy, nor does it mean a direct focus on the poorest of the poor. Rather, the opposite should be the case. Development efforts should focus on the most capable, driven and commercially oriented of the smallholder farmers and SMAEs. Assisting such farmers, e.g. in establishing marketing cooperatives, maximizes the impact of each dollar invested in terms of sustainable growth.

Perhaps as many as half of all smallholder producers are subsistence oriented (see, for example, Seville, Buxton and Vorley 2011). Such farmers are in agriculture not by choice, but because it is their survival strategy. Such farmers would benefit most from improvements in the enabling environment that facilitate their transitioning out of farming to other, more promising economic activities, including wage-earning jobs, rather than from efforts to improve their farming activities. Many of these activities and jobs can be created by upgrading and expanding food VCs (e.g. in agro-industry).

Targeting the right stakeholders also implies working with larger agribusiness or service providers, not as direct project beneficiaries but as key partners in development. Channel captains, such as large-scale commercial farms, large-scale food processors and supermarket chains, and key service providers, such as commercial banks and input manufacturers, can provide the leverage points to reach many smallholder producers or SMAEs.

Alternatively, development programmes can create an enabling environment without targeting particular stakeholders. However, in practice this means that an elite will capture the direct advantage. This is not necessarily a problem if that elite represents the most commercial smallholder farmers and their facilitated growth creates many decent jobs and strengthens the food supply, i.e. capture by the elite is merit-based, not political. Where the improved enabling environment includes better-functioning markets for land (e.g. through clear land titles and land investment regulations), subsistence smallholder farmers may derive more income from renting out their land than from farming it. Ultimately, whether directly targeted or self-selected, a core of more-entrepreneurial farmers and agribusiness service providers must be present to make SFVCD feasible.

3] Third, the strategy must target a set of upgrading activities in those parts of the VC where the largest impact in terms of growth, poverty reduction and greenness can be achieved (leverage points or root causes). Upgrading is discussed further under principle eight.

Policy and project recommendations

» Do not go from analysis to planning without first developing a realistic vision for the value chain and a core strategy for realizing that vision that (most) stakeholders in the value chain can buy into.

» Use careful targeting to ensure that value chain development strategies and plans are as uncomplicated as possible; too much complexity can hinder operational success.

» Target the stakeholders and points in the value chain or in the enabling environment that would result in the greatest impact on competitiveness and sustainability.

ILLUSTRATION OF PRINCIPLE 7
The coffee value chain in Central America

This case illustrates how a project funded by the Inter-American Development Bank used a well-targeted approach to sustainably strengthen the coffee value chain in Costa Rica, El Salvador, Guatemala, Honduras and Nicaragua after the international coffee price crisis in the early 2000s. The model implemented had three core components: access to markets, access to training and coordination and building collaboration.

The project aimed to realize a vision of sustainably integrating small- and medium-scale coffee producers into the specialty coffee market in the United States. A group of coffee buyers in the United States was identified and these firms became direct participants in the project.

The project selectively targeted cooperatives of smallholder producers who grow coffee more than 1 200 metres above sea level (a requirement for specialty coffee) that already export at least 10 percent of their production, have sound infrastructure for year-round operations, have an annual production capacity of 150 tonnes or more, are financially stable and have access to water and electricity. The coffee buyers in the United States were involved in the producer selection process and committed themselves to purchase specified amounts of coffee beans from these selected farmers as long as the set quality standard was met. By targeting a smaller and more "elite" group of small- and medium-scale producers, the project was more likely to succeed in persuading farmers to produce high-value specialty coffee.

The project then supported the selected producers with matching funds for investments in infrastructure (coffee washing stations) and technical assistance focused on one central objective: meeting the quality needed to access global markets for specialty coffee. In the process, links between coffee producers and foreign buyers were established and strengthened.

This carefully targeted support strategy was successful, bringing about sustainable economic, social and environmental improvements. Starting in 2003, the project worked with 3 000 carefully selected producers to prove that the model could work. It then increased the number of participating producers, rising to 6 000 across ten cooperatives in its last year (2009). All of these producers were able to improve their productivity and coffee quality, thus securing premium prices for larger marketed volumes. The project contributed not only to increased export volumes at the country level but also to an increased share of specialty coffee in total coffee exports; in Nicaragua, for example, this percentage jumped from 30 percent in 2003 to 50 percent in 2011. In turn, this led to increased family income, job creation, better education for children (since many of the beneficiaries spent the increased income on their children's education) and a reduced environmental footprint (as coffee waste during the wet processing stage was reduced).

Source:
Fernandez-Stark and Bamber (2012).

PRINCIPLE 8
Sustainable food value chain development is focused on upgrading

In value chain development, successful translation of a vision and strategy into an effective plan that increases competitiveness requires a realistic and complete set of carefully assessed and often innovative upgrading activities

Some form of (innovative) upgrading must take place to improve the performance (competitiveness) of a VC. This upgrading aims to achieve one or more of the triple bottom-line objectives: (1) to increase profitability by increasing efficiency and/or the value created in the end market; (2) to increase social impact by increasing inclusiveness, broadly defined; and (3) to reduce the environmental footprint of the overall chain. In today's end markets, competitiveness is increasingly determined by achieving all three objectives simultaneously. Aiming to do so increases the need for innovations that reduce the need for trade-offs between the triple bottom-line components of sustainability.

The various forms of upgrading can be classified in terms of what is being upgraded or what the upgrade aims to achieve. Classification in terms of what is being upgraded includes: technology (e.g. improved seed); organization (e.g. bulk seed purchase by a farmer group); network (e.g. contract farming linking farmers to input and output markets); and institution (e.g. an improved seed law). Classification by what the upgrade aims to achieve includes: process (e.g. introducing a food-safety protocol); product/market (e.g. from traditional markets to supermarkets); and function (e.g. farmers integrating transport to the market in their activities).

In practice, an upgrading strategy should be based on an integrated and synergistic set of individual upgrades along the VC and/or in its enabling environment. This set of upgrades must address all the critical constraints standing in the way of realizing the vision. If any of the critical factors is not addressed, the VC development effort will fail. Each individual upgrading proposed must be carefully assessed *ex ante* in terms of its anticipated impact on profitability, society and the natural environment. However, although social impact is used as an indicator of the performance of the VC, these upgrading activities are not social support programmes but rather aim to achieve sustainable broad-based improvements in competitiveness, i.e. there has to a be clear economic incentive for adopting the upgrade.

Successful upgrading implies the upgrade is adopted by a heterogeneous group of stakeholders, even if the strategy targets the more capable and commercially oriented among the smallholder farmers. This will require a flexible and diverse approach as different VC actors, e.g. young and old farmers, female and male farmers, remotely located SMAEs and SMAEs in the urban periphery, greatly differ in terms of their capacities and incentives.

Although technology leapfrogging (i.e. going straight to the latest technology and bypassing older technology, such as using cell phones to transfer money

> **ILLUSTRATION OF PRINCIPLE 8**
> **The *ndagala* value chain in Burundi**
>
> While limited in scale, this case illustrates the sustainability of impact that can be achieved through a simple but efficient technology upgrade.
>
> The *ndagala* is a sardine-like fish caught from Lake Tanganyika. The fish is dried and sold locally. It is the most important cured-fish product in Burundi.
>
> In 2004, observing that a key bottleneck in the value chain (VC) was the drying of the fish directly on the sand along the beaches, an FAO project introduced raised wire-mesh drying racks and trained producers in how to build and use them. Drying fish on the sand is unhygienic and slow, leading to significant post-harvest losses. Drying fish on the wire-mesh racks reduces drying time from three days to eight hours, allowing processors to better handle supply spikes. Because the racks are a metre from the ground, the fish is far less likely to be contaminated or eaten by insects and can be more easily covered if it rains. The technology is also less labour intensive.
>
> A review of the VC in 2013, nine years after the brief project ended, found the upgrade to be sustainable along all dimensions. Rack-dried fish sells for twice as much as sand-dried fish, while post-harvest losses are far lower and the markets that can be reached are far wider. These benefits easily offset the cost of the racks and significantly increase the incomes of the producers. Producers increased the area devoted to rack-based drying from 1 to 5 hectares between 2004 and 2013, and they continued to manage a rack-drying training centre independently. New jobs were created in processing and distribution to handle the new technology and increased volumes of dried fish. The number of people directly involved in the drying operations, mostly women, increased from 500 in 2004 to 2 000 in 2013.
>
> At the same time, the fish supplied to consumers improved in terms of taste, safety, texture, quantity and, because of the product's longer shelf-life, geographic reach into inland markets. The increase in fish supply has been achieved with little increase in amount of fish caught, and consequently little additional pressure has been placed on the lake's fish population.
>
> *Source:*
> FAO (2013c).

rather than having to visit bank branches) is often an interesting option, change typically has to be gradual because too many simultaneous changes or the skipping of critical learning steps may undermine the adoption of the upgrade.

It is thus not only the nature of the upgrade itself but also how it is delivered that has to take heterogeneity of VC actors into account. This can require the use of new and innovative ICT tools, finance products, training and educational programmes, phasing out of voucher programmes, changes in market infrastructure, information systems, extension models and so on.

The inclusion (targeting) of lead farms and firms that can champion the introduction and spread of innovative upgrades, i.e. that can drive change, is critical.

For example, if the objective is to upgrade the aggregation function of a VC, which is critical for smallholder producers, the programme must reassess the role of middlemen, possibly promoting more-coordinated, IBMs driven by current middlemen who adapt to the new situation, by entrepreneurial lead farmers or by new marketing intermediaries.

> **Policy and project recommendations**
>
> » Develop an upgrading action plan that incorporates a triple bottom-line sustainability approach but does not mix pure social support with economic development objectives.
>
> » Clearly establish the case for profitability for each proposed upgrade using realistic assumptions in terms of the likelihood of adoption and the level of impact.
>
> » Plan enough time and resources to work through the unavoidable learning processes and make sure there is a clear exit strategy in the case of a limited duration project-based approach.

PRINCIPLE 9
Sustainable food value chain development is scalable

Achieving scale, i.e. transformational change, will require that interventions focus on points of leverage or put in motion a demonstration and replication process that is based on realistic assumptions

Value-chain development is not about working with small groups of VC actors in small geographic areas ("pampered little islands of excellence"), but rather aims for impact at scale. This implies increasing the profitability of the majority of the (potentially commercial) actors in the VC; creating thousands of jobs; and increasing exports or substituting imports by double-digit percentages ("moving the needle"). In order to achieve this level of impact, facilitation and support programmes must work either through the levers in the system (e.g. policy, large-firm behaviour, institutional change or provision of business development services) or through the multiplication of a particular upgrade whose commercial viability has been proven, demonstrated and publicized and that subsequently spreads through replication. Levers and replication can also be combined to achieve scale, in that a model proven successful at the (local, small) firm level can adopted by a national-level organization (e.g. association or large agribusiness).

ILLUSTRATION OF PRINCIPLE 9
The dairy value chain in Afghanistan

The dairy value chain (VC) lends itself particularly well to both scaling up and replication. Scaling up can be driven by larger volumes of milk flowing into collection centres as a result of increased numbers of suppliers and of farmers increasing the number of milking cows they keep, both of which are encouraged by the regular cash flows associated with commercial milk production. In addition, the milk-collection-centre model can be replicated in new areas and adding value through processing creates additional income-generating opportunities. The FAO-assisted integrated-dairy-schemes approach in Afghanistan has been particularly successful in this respect.

Following the successful development of integrated dairy schemes in three areas of Afghanistan (Mazar-i Sharif, Kunduz and Kabul), the scheme was replicated a fourth time in Herat. Funded by the Government of Italy, the scheme established an interdependent set of collection points run by farmer cooperatives, a feed mill and a processing plant in Herat. Small-scale farmers with 1–5 milking cows organized themselves in village-level cooperatives. These in turn were organized into a dairy union. The union manages the milk collection system, a feed mill and a dairy plant as a vertically integrated system that links farmers directly to consumers. Value addition is the core driver of success in this model. In this, the farmers were assisted both by an increase in the capacity of the government and the private sector to deliver business services such as artificial insemination and extension, and by improved inputs for dairy farming. The final products coming out of the plant include value-added products such as fresh pasteurized milk in bottles or bags, yoghurt, cream, buttermilk and cheese.

Over 2 000 farmers, organized in 12 cooperatives, joined the scheme, which ran from 2007 to 2013. On average, farmers increased the amount of milk they delivered from 4 to 12 litres per day. The feed mill, which found markets among union members and non-members, thrived and had by the end of 2012 build up a cash reserve exceeding US$100 000. Although the completion of the dairy plant, which is essential to the overall commercial viability of the scheme, was delayed, it has since steadily increased its intake and was by 2013 operating at 60 percent of capacity (i.e. sufficient for commercial viability). Thanks to superior product quality and a loyal customer base, the scheme even survived a predatory pricing strategy by dairy-product importers, thus further demonstrating its resilience to market shocks.

Although external support ended in 2013, it is fully expected that the Herat integrated dairy scheme will follow the success of the other three schemes, which have continued their operations independently since 2010. From a social sustainability perspective, the scheme has been particularly beneficial for women, who receive and control almost 90 percent of the income from raw milk sales and used the income to pay for food, clothes, health care and education for their families.

Source:
FAO (2013d).

Scale is critical not only because a larger (positive) impact is desirable in and of itself but also because upgrading is often facilitated by a shift to a larger scale of operations. Economies of scale, reduced transaction costs and increased market power greatly enhance both the capacities and incentives that drive various upgrading processes.

There are vertical and horizontal dimensions to scale. Typically, there is an increase in scale at one level of the chain (e.g. emergence of large supermarket chains) that subsequently creates both the opportunity and the challenge to increase scale at other levels (e.g. emergence of farmer cooperatives). The VC evolves from many players conducting many small-volume transactions to few players conducting fewer but larger-volume transactions, whereby the larger scale simultaneously allows for and drives innovations and dynamics.

Policy and project recommendations

» Demonstrate how the VC development programme will achieve impact at scale along the three dimension of sustainability (and what that scale likely will be), based on realistic assumptions.

PRINCIPLE 10
Sustainable food value chain development is multilateral

Successful upgrading of a food value chain requires coordinated and collaborative efforts by the private sector, as the driver of the process, and the public sector, donors and civil society as its facilitators

Given that the overall performance of a VC is dependent on a variety of organizations, a programme to improve its performance will likely be most successful if it involves a multilateral effort with a clear differentiation in the roles played by the different stakeholders. Development approaches that expected either the public sector or the private sector to carry the burden almost unilaterally have largely failed.

In the 1960s and 1970s, government was expected to be the key driver, commonly implementing an import substitution strategy and managing marketing boards. This approach proved fiscally unsustainable, and in the 1980s structural adjustments programmes transferred the burden to the private sector.

Following the failure of the private sector to emerge as the driving force to the degree that was expected, in the 1990s non-governmental organizations came to the fore in development efforts. When the 2008 grain price crisis showed the persisting vulnerability of staple food VCs, there was a tendency for the public sector to re-engage directly in VCs and again become the driving force. It is unlikely that such efforts would be more successful today than they were in the 1960s.

There is thus a need to explore explicit multilateral approaches to overcome the deficiencies of these various unilateral approaches. The basic model that likely holds most potential is one whereby, driven by a joint vision and overall strategy, the private sector and the public sector each take the lead in specific areas. The private sector should be the driving force in increasing value creation (e.g. meeting demand for food products, creating decent jobs, increasing shareholder value and minimizing the VC's environmental footprint). The public sector, including donors and civil society, plays a facilitating and regulating role, focused on improving the business-enabling environment, e.g. laws and regulations, public infrastructure, policy, research and development.

This model implies a shift in the development support approach from short-term publicly funded projects to long-term co-funded partnerships. Rather than get directly involved in the core VC and/or impose an upgrading from the top down, the public sector takes on a facilitating role that leaves the private sector (the entrepreneur) in the driver's seat.

Facilitation can involve both more or less permanent efforts, such as selected extension or market-information services, temporary (catalytic) efforts aimed at "priming the pump," such as start-up support (e.g. loan guarantees, one-off grants and voucher schemes that gradually phase out) or the facilitation of new linkages within the private sector (neutral broker).

Even though led by the public sector, these facilitation efforts are best delivered through PPP approaches. Where the public sector is a direct participant in the VC, e.g. as a food buyer (for ministries, emergency food aid or food reserves), it can use its procurement power to facilitate upgrading activities that allow targeted actors in the food VC to become more competitive in private food markets and have a strong impact on economic, social and environmental bottom lines.

Value-chain development takes time and focuses on the creation of long-term shared value that plays out all along the VC and is expressed in the intangible characteristics of the final food product. Coordinating the efforts of the various stakeholders is, therefore, an essentially continuous effort that is greatly facilitated through the establishment of PPPs and interprofessional associations. The latter, also referred to as VC committees or commodity councils, involve stakeholders all along the VC (actors, service providers and government) cooperating in pre-competitive space.

These multilateral associations facilitate information exchange and learning related to the challenges facing all stakeholders, providing a platform for discussion and consensus-building where a joint vision and strategy can be developed among stakeholders. Established by law and mostly driven by the private sector, they can take on many additional roles, including, for example, commissioning studies, establishing industry codes of practice or standards and conducting promotional campaigns and advocacy. Advocacy also has a global component, whereby representatives of the public and private sectors join forces in dealing with the regional and global governance institutions that affect them, such as international food standards.

Phase 1: Measuring performance

Phase 2: Understanding performance

Phase 3: Improving performance

Principle 10: SUSTAINABLE FOOD VALUE CHAIN DEVELOPMENT IS MULTILATERAL

ILLUSTRATION OF PRINCIPLE 10
The salmon value chain in Chile

The emergence and resilience of Chile's salmon value chain (VC) is largely the result of strong collaboration between the public and private sector and their long-term commitment to a shared vision. However, environmental challenges remain.

Emergence
The emergence of the Chilean salmon VC can be traced back to *Fundación Chile,* a non-profit technological think tank created in 1976 by the Chilean government and the ITT Corporation. Chile's comparative advantages for salmon farming (e.g. suitable climate and extensive coastal water resources) induced *Fundación Chile* in 1982 to establish *Salmones Antárctica* as a limited company. After a long, painstaking process of trial and error, working closely with farmers, government agencies (e.g. on licensing and sanitary standards) and public research institutes (e.g. on feed formulation), *Salmones Antárctica* demonstrated the commercial viability of salmon farming in Chile. As a result, private-sector investment grew rapidly and the growth of the salmon VC took off, with impressive results. Chile became the second largest producer of farmed salmon in the world after Norway. The value of exports grew eightfold, from US$291 million in 1993 to US$2.4 billion in 2008. The ratio of value-added product (smoked salmon fillets, for example, as opposed to tailless, headless salmon carcasses) increased from 23 percent to 69 percent of total salmon industry exports between 1994 and 2004. Even though some traditional capture fishery jobs were lost as salmon farming expanded, about 45 000 (2006 estimate) new jobs offering more stable income were created in the extended VC, resulting in a strong positive impact on poverty.

Resilience
When a major infectious disease outbreak in 2007 exposed serious weaknesses in the rapidly growing salmon VC's disease control measures, a rapid response coordinated jointly by the public and private sectors ensured that such measures were quickly implemented and enforced, thus reversing the decline in harvested volumes caused by the disease. At the same time, long-term efforts involving the government, the industry and the financial sector introduced new production models, laws and regulations that greatly strengthened operational control and compliance with process standards in the VC.

Socio-environmental challenges
Whether or not the salmon VC in Chile represents a fully sustainable food VC is not clear, as environmental issues (impact on marine ecosystems) and social issues (worker conditions) have not yet been fully assessed. However, several Chilean fish farms received a Best Aquaculture Practices certification in 2013, which appears to indicate that these issues are being addressed.

Sources:
UNCTAD (2006); Alvial *et al.* (2012); Niklitschek *et al.* (2013); SeafoodSource.com (accessed July 2013).

4 » Ten principles in sustainable food value chain development

Policy and project recommendations

» Recognize the complementary synergy-creating roles played by the public sector, the private sector and civil society in the upgrading food value chains and facilitate the emergence of a joint vision and strategy.

» Facilitate the emergence of continuous partnerships between the public and private sectors and civil society.

Phase 1: Measuring performance

Phase 2: Understanding performance

Phase 3: Improving performance

Principle 10:
SUSTAINABLE FOOD VALUE CHAIN DEVELOPMENT IS MULTILATERAL

CHAPTER 5
Potential and limitations

The VC development approach reflects the cumulative and progressive outcome of development thinking and learning from practice over the past 30 years. It provides a sufficiently broad and flexible framework to make it relevant for and adaptable to many economic development challenges. VC development's focus on addressing root problems in underperforming systems ensures that the resulting development strategies and plans (policies and support programmes) have the potential to efficiently address poverty and food security in a significant and sustainable way.

Given the prevalence of the approach, however, it surprising how little critical reflection there has been on whether VC development has had a greater impact than some alternate approaches (e.g. those focused on solving specific problems outside of a systems context, or those that blend social and developmental objectives).

Part of the problem is that in VC development it is often difficult to link outcomes and impacts such as poverty reduction to activities and to assess the scale and sustainability of the outcomes. This is due to the inherent complexity of VCs and to the fact that impact often occurs, and hence can only be measured, after support programmes end. Measuring impact would thus require further funding. This increases the cost and therefore reduces the number of impact assessments performed. For example, in a study of 30 donor-funded VC projects, Humphrey and Navas-Aleman (2010) found little independent systematic evaluation of impact beyond simple project activity and output verification. Beyond this, assessing the developmental impact of VC interventions appears to quickly enter the realm of anecdotal evidence and wishful thinking. This is a weakness of VC development that has yet to be addressed.

Value-chain development in practice is not without its challenges and limitations. These include the following:

» First, there is no common understanding of the VC development concept or agreement on how to implement it. For example, in a critical review of the VC approach within the United Nations system, Stamm and von Drachenfels (2011) find that member agencies do not have clear definitions of the VC development paradigm that are well-communicated internally. This in turn undermines external transparency.

As a result of the lack of a universal understanding, many development efforts are fashionably branded with the VC label, yet violate one or more of the principles of VC development as defined here. They do not, for example, address root problems, do not start from a clear market opportunity for creating added value or do not target farms and agribusiness that have the potential to be commercially viable (but rather focus on pure subsistence farming). Such mislabelled efforts often involve much direct intervention by the public sector or they critically depend on public support without a clear or realistic exit strategy.

5 » Potential and limitations

» Second, the emphasis in VC development is often still on the economic and financial aspects, with social and environmental impacts being considered only peripherally if at all. Even though this is now changing, a conscious effort is required to ensure the sustainability of the upgraded VC in social and environmental terms. The risk that then could emerge, especially in the absence of a common understanding, is that VC development is confused with social support or environmental protection programmes which are of a fundamentally different nature.

» Third, VC development is complex and time-consuming, and taking short cuts comes at a price. In practice, however, time and other resources are often insufficient to holistically assess the complex VC system, resulting in flawed designs for development projects and programmes. Trust and learning, key ingredients in VC development, do not emerge overnight.

Economic activity, especially in the agrifood sector, is a cyclical process that takes time and is exposed to external shocks that can lead to setbacks, e.g. drought, social unrest and political change. That development of food VCs is typically supported through rigidly structured, short-duration (3–5 year) projects worsens the problem at an operational level and points to long-term partnership approaches as perhaps a better way forward in SFVCD.

» Fourth, VC development is a fragmented, "one-chain-at-a-time" process that has three well-recognized blind spots:

> **VALUE CHAIN DEVELOPMENT HAS GREAT POTENTIAL BUT ALSO LIMITATIONS**

1] Decision-making by actors: Many development efforts targeting food VCs do not achieve sustainable impact because they are too narrowly focused on the commodity at hand. This is especially relevant at the farm level. Farmers are typically not specialized in a single crop (e.g. because of crop rotation) or even in crop production (combining it with livestock or fisheries), and their decisions in one particular VC depend on their decisions in other VCs. To change a farmer's behaviour in a particular VC (e.g. to get them to sell maize into a particular market) there may well be a need to also promote change in another VC (e.g. to provide a market opportunity for soybeans). In other words, there is a need to better integrate farming systems thinking into efforts to develop food VCs.

2] VC development versus food system development: Achieving broad-based developmental impact by facilitating growth in a particular VC requires taking a broader look at interactions of all food VCs at the food-system level. For example, the mechanization of maize production may result in more jobs being lost at the farming level than it creates at other links in the VC. As another example, developing the palm oil VC to supply bioenergy markets may take land away from food production, possibly driving up food prices and undermining overall food security. As these are socially undesirable outcomes, there is a need for careful selection of VCs to target and for complementary programmes that allow for the anticipated negative impacts of the development of a particular food VC to be offset by other simultaneous developments in the food system and beyond (development of non-food VCs, self-employment, trade and so on).

3] Synergies across VCs: Making food VCs perform well from a sustainability perspective depends in large measure on their interdependencies. Examples here include: aggregating the demand from various food VCs may create the critical mass that makes the provision of certain services or inputs commercially or fiscally viable; linking VCs that produce raw agricultural commodities at different times of year may provide the year-round supply of materials to be processed that makes certain types of processing economically feasible; clustering comparable links of different food VCs (e.g. in a food processing zone) may stimulate learning and create an important leverage point for the food system as a whole; and having markets for by-products from a particular food VC in other (food or non-food) VCs may greatly affect profitability in the supported VC. There is a need to create and exploit these synergies as much as possible.

Potential and limitations

In summary, some issues go beyond the scope of VC concept. These include: assisting those households and SMAEs that cannot be realistically included as the VC becomes more competitive; the effective and efficient delivery of public goods and services that are not commodity specific; the role of nutrition in consumer health; and the management of natural resources and food security at the national level. These various limitations of the VC development approach highlight the need for the public sector to engage in broad-based national development programmes, transitional strategies, safety nets and other social support mechanisms, nutritional awareness campaigns and targeted environmental programmes that not only complement VC development efforts but even guide them.

FAO's SFVCD handbooks, of which this booklet is the first being published, aims to start addressing these challenges. It offers some clarity and practical advice to facilitate better formulated policies, better designed and implemented projects and programmes and more and higher-quality assessments of the sustainability of the impact of SFVCD.

CHAPTER 6
Conclusions

The ultimate objective of SFVCD is to contribute significantly to a broad-based improvement in the welfare of a society, for both the current and future generations. For the specific situation of VCs in the food system, this publication presents a concept, an analytical framework, a development paradigm and a set of ten principles that explicitly incorporate the multidimensional nature of the concepts of value added and sustainability.

The value added in VCs is captured in five ways: returns to asset owners; wage incomes; benefits to consumers; tax revenues; and impacts on the environment, broadly defined. This breakdown of the value-added concept allows for performance assessments that go beyond competitiveness and inclusion of smallholder farmers. Rather, impact is simultaneously assessed against the three dimensions of sustainability: economic, social and environmental. Broad-based wealth accumulation, the number and nature of direct and indirect jobs created, an improved food supply, a strengthened tax base and a lighter environmental footprint of food production and distribution all contribute to the performance of a food VC.

Value chain development cannot solve all problems in the food system. Food VCs cannot provide incomes for everyone, cannot incorporate trade-offs at the food-system level and cannot entirely avoid negative environmental impacts. Public programmes and national development strategies are needed to address these limitations. However, such programmes and strategies are largely financed through tax revenues generated by VCs, thus placing VC development in general, and SFVCD in particular, at the heart of any strategy aimed at reducing poverty and hunger in the long run.

Conclusions

References

Alvial, A., Kibenge, F., Forster, J., Burgos, J.M., Ibarra, R. & St-Hilaire, S. 2012. *The recovery of the Chilean salmon industry: The ISA crisis and its consequences and lessons.* St. Louis, MO, USA, The Global Aquaculture Alliance (available at http://www.gaalliance.org/cmsAdmin/uploads/GAA_ISA-Report.pdf).

Bain, J.S. 1956. *Barriers to new competition.* Cambridge, MA, USA, Harvard University Press.

Barrett, C.B. 2008. Smallholder market participation: Concepts and evidence from Eastern and Southern Africa. *Food Policy*, 33(4): 299–317.

Berry, R.A. & Cline, W.R. 1979. *Agrarian structure and productivity in developing countries.* Baltimore, MD, USA, The Johns Hopkins University Press.

Blanchard, D. 2010. *Supply chain management best practices.* 2nd ed. Hoboken, NJ, USA, Wiley.

Blue Skies. 2010. *Making fruit happy. Blue Skies Sustainability Report 2008/2009.* Pitsford, Northamptonshire, UK, Blue Skies Holdings (available at http://www.blueskies.com/happyfruit.pdf).

Blue Skies. 2012. *The JEE Report 2010–2011. Our blue print for a sustainable business.* Pitsford, Northamptonshire, UK, Blue Skies Holdings (available at http://www.blueskies.com/jee.pdf).

Bowersox, D., Closs, D., Cooper, M. & Bowersox, J. 2013. *Supply chain logistics management.* 4th ed. New York, USA, McGraw-Hill.

Carter, M.R. 1984. Identification of the inverse relationship between farm size and productivity: An empirical analysis of peasant agricultural production. *Oxford Econ. Pap.*, 36: 131–145.

CPDA. 2008. *Report on the small-scale tea sector in Kenya.* Nairobi, Christian Partners Development Agency (available at http://somo.nl/publications-en/Publication_3097).

Concepcion, S.D., Digal, L. & Uy, J. 2007. *Innovative practice Philippines: The case of NorminVeggies in the Philippines.* Regoverning Markets Innovative Practice series. London, International Institute for Environment and Development (available at http://pubs.iied.org/pdfs/G03259.pdf).

Cornia, G.A. 1985. Farm size, land yields and the agricultural production function: An analysis for fifteen developing countries. *World Dev.*, 13(4): 513–534.

Costello, C., Demont, M. & Ndour, M. 2013. Marketing local rice to African consumers. *Rural 21*, 47(1): 32–34.

Demont, M. 2013. Reversing urban bias in African rice markets: A review of 19 national rice development strategies. *Global Food Secur.*, 2(3): 172–181.

Demont, M. & Rizzotto, A.C. 2012. Policy sequencing and the development of rice value chains in Senegal. *Dev. Pol. Rev.*, 30(4): 451–472.

Donovan, J., Cunha, M., Franzel, S., Gyau, A. & Mithöfer, D. 2013. *Guides for value chain development – a comparative review.* Wageningen, The Netherlands, CTA, and Nairobi, World Agroforestry Centre.

Ericksen, P.J. 2008. Conceptualizing food systems for global environmental change research. *Global Environ. Chang.*, 18(1): 234–245.

FAO. 2006. *Food security.* Policy Brief, Issue 2. Rome.

FAO. 2009. *The potato supply chain to PepsiCo's Frito Lay in India*, by M. Punjabi. Unpublished report. Rome.

Note: All internet links correct as of 30 January 2014.

FAO. 2012. *Smallholder business models for agribusiness-led development: Good practice and policy guidance*, by S. Kelly. Rome.

FAO. 2013a. *Organic agriculture: African experiences in resilience and sustainability*, edited by R. Auerbach, G. Rundgren & N. El-Hage Scialabba. Rome.

FAO. 2013b. *Making Kenya's efficient tea markets more inclusive.* MAFAP Policy Brief #5. Rome (available at http://www.fao.org/docrep/018/aq657e/aq657e.pdf).

FAO. 2013c. *Simple fish-drying racks improve livelihoods and nutrition in Burundi.* Rome.

FAO. 2013d. *Development of integrated dairy schemes in Herat.* GCP/AFG/046/ITA. Report of Final Review Mission 9–25 December 2012. Rome, Office of Evaluation, FAO.

Feller, A., Shunk, D. & Callarman, T. 2006. *Value chains vs. supply chains.* BPTrends, March 2006 (available at http://www.ceibs.edu/knowledge/papers/images/20060317/2847.pdf).

Fernandez-Stark, K. & Bamber, P. 2012. *Assessment of five high-value agriculture inclusive business projects sponsored by the Inter-American Development Bank in Latin America.* Durham, NC, USA, Center on Globalization, Governance and Competitiveness, Duke University.

Gereffi, G. & Korzeniewicz, M. (eds). 1994. *Commodity chains and global capitalism.* Westport, CT, USA, Praeger, pp. 95–122.

Gereffi, G., Humphrey, J. & Sturgeon, T. 2005. The governance of global value chains. *Rev. Int. Polit. Econ.*,12(1): 78–104.

GIZ. 2011. *Financing agricultural value chains in africa – a synthesis of four country case studies.* Bonn, Germany.

Gómez, M.I., Barrett, C.B., Buck, L.E., De Groote, H., Ferris, S., Gao, H.O., McCullough, E., Miller, D.D., Outhred, H., Pell, A.N., Reardon, T., Retnanestri, M., Ruben, R., Struebi, P., Swinnen, J., Touesnard, M.A., Weinberger, K., Keatinge, J.D.H., Milstein, M.B. & Yang, R.B. 2011. Research principles for developing country food value chains. *Science* 332(6034): 1154–1155.

Haggblade, S.J. & Gamser, M.S. 1991. *A field manual for subsector practioners.* Bethesda, MD, USA, GEMINI (available at http://www.microlinks.org/sites/microlinks/files/resource/files/A%20Field%20Manual%20for%20Subsector%20Practioners.pdf).

Haussman, R., Rodrik, D. & Velasco, A. 2005. Growth diagnostics. *In* N. Serra & J.E. Stiglitz, eds. *The Washington Consensus reconsidered*, pp. 324–354. Oxford, UK, Oxford University Press.

Heltberg, R. 1998. Rural market imperfections and the farm size–productivity relationship: Evidence from Pakistan. *World Dev.*, 26(10): 1807–1826.

Hobbs, J.E., Cooney, A. & Fulton, M. 2000. *Value chains in the agri-food sector.* Saskatoon, Saskatchewan, Canada, College of Agriculture, University of Saskatchewan.

Humphrey, J. & Navas-Aleman, L. 2010. *Value chains, donor interventions and poverty reduction: A review of donor practice.* Brighton, UK, Institute for Development Studies.

Ikerd, J. 2011. Essential principles of sustainable food value chains. *J. Agr. Food Syst. Community Dev.*, 1(4): 15–17.

ILO. 2007. *Toolkit for mainstreaming employment and decent work.* Geneva, Switzerland, International Labour Organization.

ILO. 2011. *Social protection floor for a fair and inclusive globalization.* Geneva, Switzerland, International Labour Organization.

Jackman, D. & Breeze, J. 2010. *A guide to inclusive business.* London, The International Business Leaders Forum.

Kaplinsky, R. & Morris, M. 2000. *A handbook for value chain research.* Ottawa, International Development Research Center.

Knopp, D. & Foster, J. 2010. *The economics of sustainability.* The Woods Family Trust and the Gatsby Trust (available at http://www.idhsustainabletrade.com/site/getfile.php?id=331).

Kubzansky, M., Cooper, A. & Barbary, V. 2011. *Promise and progress – market-based solutions to poverty in Africa.* Mumbai, India, The Monitor Group (available at http://www.mim.monitor.com/downloads/PromiseAndProgress-Full-screen.pdf).

Lauret, F. 1983. Sur les études de filières agroalimentaires. *Écon. et Soc.,* 17(5): 721–740.

Lazzarini, S.G., Chaddad, F.R., & Cook, M.L. 2001. Integrating supply chain and network analyses: The study of netchains. *J. Chain Netw. Sci.,* 1(1): 7–22.

Lee, R.G., Flamm, R., Turner, M.G., Bledsoe, C., Chandler, P., De Ferrari, C., Gottfried, R., Naiman, R.J., Schumaker, N., & Wear, D. 1992. Integrating sustainable development and environmental vitality: A landscape ecology approach. *In* R.J. Naiman, ed. *Watershed management: Balancing sustainability and environmental change,* pp. 499–521. New York, USA, Springer-Verlag.

Lundy, M., Becx, G., Zamierowski, N., Amrein, A., Hurtado, J.J., Mosquera, E.E., & Rodríguez, F. 2012. *LINK methodology: A participatory guide to business models that link smallholders to markets.* Publication No. 380. Cali, Colombia, Centro Internacional de Agricultura Tropical.

Moustier, P. & Leplaideur, A. 1999. Cadre d'analyse des acteurs du commerce vivrier africain. Serie Urbanisation, alimentation et filières vivrières No. 4. Montpellier, France, CIRAD. 42 pp.

Neven, D. 2009. *Three steps in value chain analysis.* microNote No. 53. Washington, DC, United States Agency for International Development (available at http://www.microlinks.org/sites/microlinks/files/resource/files/mn_53_three_steps_in_vc_analysis.pdf).

Niklitschek, E.J., Soto, D., Lafon, A., Molinet, C. & Toledo, P. 2013. Southward expansion of the Chilean salmon industry in the Patagonian Fjords: main environmental challenges. *Rev. Aquacult.,* 5(3): 172–195.

Porter, M.E. 1985. *Competitive advantage.* New York, The Free Press.

Porter, M.E. & Kramer, M.R. 2011. Creating shared value. *Harvard Bus. Rev.,* 89(1/2): 62–77.

Reardon, T. & Timmer, C.P. 2012. The economics of the food system revolution. *Annu. Rev. Resour. Ec.,* 4: 14.11–14.40.

Reardon, T., Chan, K., Minten, B. & Adriano, L. 2012. *The quiet revolution in staple food value chains: Enter the dragon, the elephant and the tiger.* Manila, Asian Development Bank, and Washington, DC, International Institute for Food Policy Research.

Sayer, J., Sunderland, T., Ghazoul, J., Pfund, J.-L., Sheil, D., Meijaard, E., Venter, M., Klintuni Boedhihartono, M., Day, M., Garcia, C., van Oosten, C. & Buck, L.E. 2013. Ten principles for a landscape approach to reconciling agriculture, conservation, and other competing land uses. *Proc. Natl. Acad. Sci. USA* 110(21): 8345–8348 (available at http://www.pnas.org/content/early/2013/05/14/1210595110.abstract).

Seville, D., Buxton, A. & Vorley, B. 2011. *Under what conditions are value chains effective tools for pro-poor development?* Hartland, VT, USA, Sustainable Food Lab, and London, International Institute for Environment and Development (available at http://pubs.iied.org/16029IIED.html).

Staatz, J.M. 1997. *Notes on the use of subsector analysis as a diagnostic tool for linking industry and agriculture.* Staff Paper 97-4. East Lansing, MI, USA, Department of Agricultural Economics, Michigan State University.

Stamm, A. & von Drachenfels, C. 2011. *Value chain development: Approaches and activities by seven UN agencies and opportunities for interagency cooperation.* Geneva, Switzerland, International Labour Organization.

Sun Star. 2011a. *Northern Mindanao group cites as market facilitator for small farmers.* (available at http://www.sunstar.com.ph/cagayan-de-oro/business/northern-mindanao-group-cited-market-facilitator-small-farmers).

Sun Star. 2011b. *The NorMinVeggies experience: Finding strength in consolidation.* (available at http://www.sunstar.com.ph/cagayan-de-oro/business/2011/11/16/norminveggies-experience-finding-strength-consolidation-190960).

The Hindu Business Line. 2012. *More Bengal farmers turn to Atlanta potatoes.* (available at http://www.thehindubusinessline.com/industry-and-economy/agri-biz/more-bengal-farmers-turn-to-atlanta-potatoes/article4203936.ece).

UNCTAD. 2006. *Transfer of technology for successful integration into the global economy: A case study of the salmon industry in Chile.* New York, USA, United Nations Conference on Trade and Development (available at http://unctad.org/en/Docs/iteiit200512_en.pdf).

USAID. 2009. *Global food security response: West Africa rice value chain analysis.* MicroReport #161. Washington, DC (available at http://www.microlinks.org/library/global-food-security-response-west-africa-rice-value-chain-analysis).

van Engelen, A., Malope, P., Keyser, J. & Neven, D. 2012. *Botswana agricultural value chain project: Beef value chain study.* Rome, FAO, and Gaborone, Ministry of Agriculture, Botswana.

Webber, M. 2007. *Using value chain approaches in agribusiness and agriculture in sub-Saharan Africa: A methodological guide.* Report prepared for the World Bank by J.E. Austin Associates, Inc (available at http://www.technoserve.org/files/downloads/vcguidenov12-2007.pdf).

Wiggins, S. & Keats, S. 2013. *Leaping and learning: Linking smallholders to markets.* London, Agriculture for Impact, Imperial College London (available at https://workspace.imperial.ac.uk/africanagriculturaldevelopment/Public/LeapingandLearning_FINAL.pdf).

World Bank. 2009. *World development report 2009: Reshaping economic geography.* Washington, DC.

World Bank. 2013. *World development report 2013: Jobs.* Washington, DC.

ANNEX
Concepts related to the value-chain concept

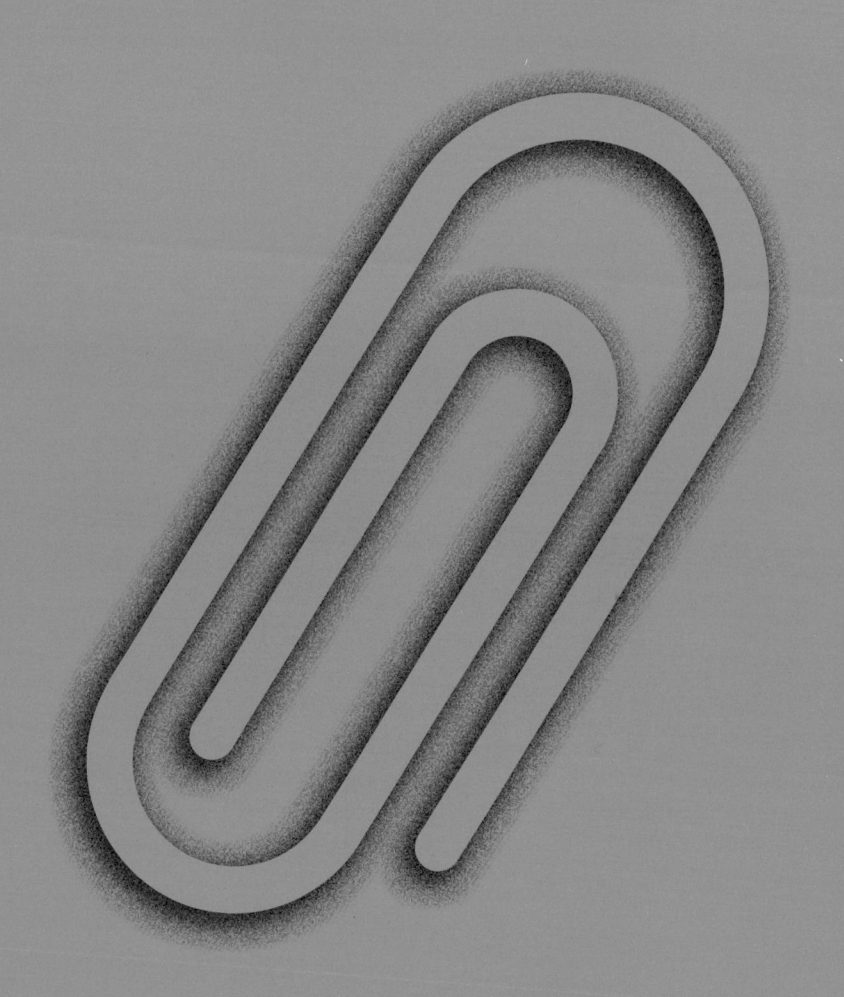

Several concepts that are related to the VC concept are described briefly in this annex. The years in parentheses indicate when these terms began to be used in economic development literature. Although these terms are often used interchangeably, they represent different notions.

Filière/commodity chain (1950s)

The *filière* (or commodity chain) approach historically focused on linking production systems to large-scale processing and final consumption from a mostly technical perspective. A *filière* maps and quantifies physical product flows from one actor to the next and assesses aspects such as transport and storage logistics and technical conversion ratios in product handling and processing. The approach has its origins in the former French colonies, where it was used to improve export chains for commodities such as coffee, cocoa and cotton. Since the 1980s the approach has been broadened by including generation and distribution of income among actors, as well a behavioural model for actors (incentives and capacities), collective action, market power, overall chain governance, including sectoral organization and institutions, and spillover effects on the wider economy. As such, the concept has become similar to the VC concept. Key references are Lauret (1983) and Moustier and Leplaideur (1999).

Subsector (1970s–1980s)

A (food-based) subsector approach usually starts from a particular agricultural raw material (e.g. maize) and maps out, quantifies and analyses the various competing channels through which this material is transformed into intermediate and final products that are sold into their various markets. The concept of competing channels, each defined by particular technologies and trading relationships, allows for a deeper understanding of the competitive changes within the subsector than is offered by a filière approach. The subsector is seen as a dynamic system in which the heterogeneity of economic actors and their position in the various channels are recognized. By taking a view of the entire subsector, the location of the actors (especially micro and small enterprises) within the subsector and the relationships between the actors, the approach identifies points of leverage in order to derive cost-effective and inclusive development strategies. As such, the subsector approach is a direct precursor of the VC concept but lacks the latter's explicit treatment of the elements of governance, globalization and end-market focus. Key references are Haggblade and Gamser (1991) and Staatz (1997).

Supply chain (1980s)

Supply chains are multifirm collaborative arrangements designed to create value through integrated effort by accomplishing five critical flows: product, service, information, finance and knowledge.[19] Logistics is the primary conduit of product and service flows in the supply chain, spanning the domain from the

---

[19] Knowledge here refers to the ability to use information in a practical sense.

original production of the raw materials to the presentation of the final products in a retail outlet. It includes aspects such as packaging, information systems, equipment and facilities capacity, transport, storage, regulations and insurance. Supply chains can be assessed at the level of the individual firm (procurement, conversion and distribution) and at the overall chain level (e.g. traceability systems). The trends of globalization and industrialization have greatly increased the opportunities and challenges in supply-chain management, which emerged as a field of practice in the 1980s. Feller, Shunk and Callarman (2006), Blanchard (2010) and Bowersox *et al.* (2013) provide good introductions to the concept.

Porter's value chain (1985)

Unlike the VC concept as presented in this publication, Porter's VC concept is a firm-level concept (Porter 1985). In particular, it facilitates the systematic assessment of what unique characteristics a firm has or can develop to create competitive advantages that allow it to profitably sell a similar quality product for less or to sell a differentiated product for more than its competitors. The increased value that is created is shared between the firm (profit) and the consumer (satisfaction or savings). Competitive advantage, and thus value-creation opportunities, can be found or created through five primary activities (inbound logistics, outbound logistics, operations, marketing and customer service) and four support activities (firm infrastructure, human resources management, technology development and procurement). As such, Porter's VC is a business strategy tool, the main objective of which is to help managers decide how to profitably increase the competitiveness of the firm. It does not assess value added at the level of the entire chain. Porter's VC concept was recently expanded to incorporate the shared-value paradigm, which takes a broader and more long-term perspective on creation of competitive value (Porter and Kramer 2011). Specifically, it incorporates the value that is created at other points in the VC, and especially for society overall, which both strengthens critical supplier–buyer linkages and creates value for consumers. The shared-value concept brings the two VC concepts closer together, especially in terms of sustainability, even though assessment of the competitiveness of the individual firm and facilitating managerial decision-making remain the central objectives of Porter's shared-value concept.

Global commodity chain (1994)

The global-commodity-chain concept combines the concepts of value added and globalization. It emphasizes the growing importance of global firms (retailers and brand marketers) and how they coordinate the activities of the various firms in production and distribution networks that stretch across multiple countries (Gereffi and Korzeniewicz 1994). As such, the concept highlights the importance of understanding final consumer markets as key drivers of VC dynamics. Governance of a VC is seen as being influenced by three main factors: (1) the complexity of the information needed to coordinate transactions along the chain; (2) how easily the transaction information can be codified (e.g. through standards); and (3) how capable suppliers are of meeting the transaction requirements (Gereffi, Humphrey and Sturgeon 2005).

Net-chain (2001)

The net-chain concept merges the concepts of the supply chain and the network of a firm. It is defined as a set of vertically layered networks of horizontal ties within an industry (Lazzarini, Chaddad and Cook 2001). The main focus is on interorganizational collaboration and its impact on coordination, quality management and, ultimately, value creation. As such it relates mainly to the vertical and horizontal linkages in VCs, with added value derived from an improved (optimized) architecture along both dimensions: governance along the vertical axis, collective action along the horizontal axis and actor–support-provider linkages along both axes. Uptake of the concept has been limited in the economic development field.

(Inclusive) business model (2005)

The business model is a narrower concept than the VC. It is mostly seen at the level of the individual firm and how it approaches value capture and growth (e.g. franchising and ownership are two different business models for retail expansion). In economic development, it is used to study the nature of a particular link in the VC. Most notably in food chains, the focus is on the critical and often chain-wide weakest link between smallholder producers and their direct buyers. The ongoing revolution in the food system is pushing for ever-increasing levels of coordination in chains for both staple food and high-value foods. This makes working through traditional middlemen who purchase raw agricultural materials via unplanned spot-market transactions an inadequate business model. Rather, it calls for the development of new models with greater coordination capacity in which either traditional middlemen take on new roles or where new types of marketing intermediaries emerge (e.g. lead farmers, specialized new entrants and marketing cooperatives). By incorporating new financing, knowledge-sharing, input access and output marketing approaches, these innovative models may allow for the inclusion of large numbers of smallholder producers, in which case they are referred to as inclusive business models (IBMs). Another important link from a development perspective is that between a food processor and poor consumers, in which innovation in products and distribution model (e.g. fortified foods and new retailer networks) can bring healthy foods within reach of poor consumers (IBMs of the bottom-of-the-pyramid type). A key driver behind the growing importance of the business-model approach in both research and real-world application is that the business model is a much more manageable and quickly implemented concept than the VC approach, which includes far more elements (all actors, all channels and all environmental elements). However, even though the business model focuses on the specific components of a specific link in the chain, it will still, as in VC development, search for root causes of underperformance and for elements of upgrading strategy wherever they may be located in the VC or its environment. Jackman and Breeze (2010), Kubzansky, Cooper and Barbary (2011), FAO (2012) and Lundy et al. (2012) provide good introductions to IBMs.

Food system (2008)

The food system is a broader concept than the food VC, involving all processes and infrastructure required to feed a population. It comprises all food VCs that affect a selected set of food markets (e.g. those in a particular country). As such, the dimensions it adds are the synergies that are created by developing common elements across various VCs, be they non-VC-specific service providers (such as logistics firms), elements of the enabling environment (such as land-title laws) or links between different food chains (such as a by-product in one chain being an input in another chain). Food systems can have various subsystems – global or local, conventional or organic, large-scale or niche and so on. The food system also adds an overall societal perspective, including food security, health, nutrition, employment, research, education, protection of the natural environmental, trade-offs between the multiple objectives of agriculture (food, feed, fuel and fibre) and so on. It thus has political, economic, social and environmental dimensions. An analysis of the overall food system is an important step in VC development, especially in terms of VC selection and maximizing the impact of public support programmes. More on the food systems concept can be found in Ericksen (2008) and in Reardon and Timmer (2012).

Landscape system (2010)

The landscape-system approach combines geographical, natural and socio-economic elements to tackle economic, social and environmental challenges related, in particular, to the use of natural resources (ecosystem preservation). The object is to develop a deep understanding of how multiple uses of natural resources (land, water, plants and animals, air, etc.) are interrelated in a given location, based on which strategies can be designed that are more likely to simultaneously increase food production, improve household welfare and reduce the environmental footprint. The approach is not new, but is gaining importance. An elaboration on this approach can be found in Lee *et al.* (1992) and Sayer *et al.* (2013).